Principles of AC Machines

This book offers a clear and concise introduction to the theory and applications of alternating current (AC) electrical machines. It explores electromechanical systems, focusing on the basic concepts of electromagnetic fields and magnetic materials, delving into the governing laws and forces of magnetism. The text then transitions to a comprehensive examination of electromechanical AC machines, including AC generators, transformers, and synchronous and induction machines, with particular attention to their operation, analysis, and design. It includes a wealth of solved examples that elucidate complex principles and unsolved exercises to strengthen problem-solving prowess.

Features:

- Offers in-depth information and study about asynchronous and synchronous machines
- Presents complex electromechanical principles in an easy-to-follow manner
- Clarifies important ideas in electromechanics by offering several solved and unsolved examples
- Covers over 150 practical examples, solved problems, and exercises to reinforce learning

This book is designed for undergraduate and beginning graduate students in electrical engineering, renewable energy engineering, and related disciplines. It is also a practical reference for professionals working in the power systems, automation, and energy sectors.

Pathways to a Sustainable Energy Future: Innovations and Strategies for a Low-Carbon World

Series Editor: Ramesh C. Bansal

Principles of AC Machines
Ahmed G. Abokhalil and Ramesh C. Bansal

Principles of AC Machines

Ahmed G. Abokhalil and
Ramesh C. Bansal

CRC Press
Taylor & Francis Group
Boca Raton London New York

CRC Press is an imprint of the
Taylor & Francis Group, an **informa** business

Designed cover image: Shutterstock

First edition published 2027
by CRC Press
2385 NW Executive Center Drive, Suite 320, Boca Raton FL 33431

and by CRC Press
4 Park Square, Milton Park, Abingdon, Oxon, OX14 4RN

CRC Press is an imprint of Taylor & Francis Group, LLC

ISBN: 978-1-032-85318-5 (hbk)
ISBN: 978-1-041-35058-3 (pbk)
ISBN: 978-1-003-79584-1 (ebk)

DOI: 10.1201/9781003795841

Typeset in Times
by Apex CoVantage, LLC

Contents

Preface

The modern world depends heavily on electrical energy, and at the heart of its efficient generation, transmission, and utilization lies the comprehensive understanding of electrical machines. This textbook, *Principles of AC Machines*, is crafted to serve as a foundational resource for undergraduate and early postgraduate students pursuing electrical, electronic, or renewable energy engineering. It is equally valuable for professionals seeking to refresh their knowledge in this crucial domain.

This book begins by introducing the fundamental concepts of electromagnetic theory, including magnetic circuits, flux behavior, hysteresis, eddy current losses, and energy conversion principles. These principles are essential for understanding the operation of both rotating and static electrical machines.

In the subsequent chapters, this book explores the operation and analysis of transformers and AC rotating machines – specifically synchronous, three phase and single phase induction machines, and stepper motors. Each topic is presented with clear theoretical explanations, practical examples, mathematical derivations, and problem-solving exercises. Special emphasis is placed on magnetic materials, sinusoidal excitation, and the behavior of rotating magnetic fields, laying the groundwork for more advanced machine modeling and control topics.

This book also integrates numerous solved examples and exercises to reinforce learning and aid in the development of analytical skills.

We express our sincere gratitude to the institutions, colleagues, and students who contributed to the development and refinement of this work. We hope that this book will inspire a deeper interest in the field of electric machines and serve as a stepping stone for advanced study and innovation.

About the Authors

Ahmed G. Abokhalil received his PhD from Yeungnam University, South Korea. Before joining the faculty at the University of Sharjah in 2021, he was an associate professor at Majmaah University, Saudi Arabia, where he worked for eight years. Prior to that, he was an assistant professor at Assiut University, Egypt, from 2010 to 2012. He is currently an associate professor in the SREE Department at the University of Sharjah. Abokhalil's research is in the general area of renewable energy, power electronics, power systems, and modeling and control of renewable energy systems. Dr. Abokhalil has received several internal and external funding, and he has authored and coauthored more than 100 publications.

Ramesh C. Bansal has more than 25 years of diversified experience in research, scholarship of teaching and learning, accreditation, industrial, and academic leadership in several countries. At present, he is a professor in the Department of Electrical and Computer Engineering at the University of Sharjah. Previously, he was a professor and the group head (power) in the ECE Department at the University of Pretoria (UP), South Africa. Prior to his appointment at UP, he was employed by the University of Queensland, Australia; the University of the South Pacific, Fiji; BITS Pilani, India; and the Civil Construction Wing, All India Radio.

Professor Bansal has significant experience of collaborating with industry and government organizations. These utilities include NTPC, Powerlink, and ESKOM. During sabbaticals, he has worked with Powerlink, an Australian government-owned corporation responsible for Queensland's high-voltage electricity transmission network. He has made significant contributions to the development and delivery of BS and ME programs for utilities. He has extensive experience in the design and delivery of CPD programs for professional engineers. He has carried out research and consultancy and attracted significant funding from industry and government organizations.

Professor Bansal has published over 500 journal articles, presented papers at conferences, books, and chapters in books. He has Google citations of over 25,000 and an h-index of 75. He has supervised 25 PhD and 10 postdocs/research assistants and is currently supervising several PhD students. His diversified research interests are in the areas of renewable energy (wind, PV, DG, microgrids) and smart grids. Professor Bansal is an editor of several highly regarded journals, such as *IEEE Systems Journals*, *Electric Power Components and Systems*, and *Smart Grids and Sustainable Energy*. He is a fellow and chartered engineer of IET-UK, a fellow of the Institution of Engineers (India), a fellow of the SAIEE, South Africa, and a senior member of the IEEE-USA.

1 Introduction to Electrical Machines

1.1 INTRODUCTION

An electrical machine converts mechanical energy into electric energy, converts electric energy into mechanical energy, or transforms electric energy by changing the type of current or by altering such parameters as voltage and frequency. The operation of electric machines is based on electromagnetic induction and the laws governing the interaction of electric currents and magnetic fields. This chapter presents an introduction to the basics of electrical machines.

1.2 MAGNETIC CIRCUIT ANALYSIS

A magnetic circuit is described as the closed route that magnetic flux follows. Similar to how an electric circuit facilitates the flow of electric current, a magnetic circuit facilitates the flow of magnetic flux. Understanding magnetic circuits is critical in the creation, analysis, and use of electromagnetic devices, including transformers, electric rotating machines, and electromagnetic relays. The fundamental principles of magnetic circuits are outlined in subsequent sections.

1.2.1 Electromagnetism

The early nineteenth century saw the first observations and measurements of electromagnetic phenomena, notably by three European scientists: Hans Christian Oersted, Andre Marie Ampere, and Michael Faraday. Oersted discovered that a current-carrying wire could cause a nearby suspended magnet to move. In 1825, Ampere showed that a wire with current could exert a force on another wire with current flowing in the reverse direction. In 1831, Faraday found that moving a magnet close to a wire coil could induce current flow in the coil. These experiments laid the foundation for all electromechanical devices, especially motors and generators. The early experiments confirmed that an electrical current traveling through a wire generates a magnetic field around it, as depicted in Figure 1.1a. This magnetic field can be visualized by displaying lines of magnetic flux, symbolized by ϕ. The direction of this field can be determined using the "right hand rule," where one imagines wrapping their right hand around a wire with the thumb pointing towards the direction of current flow, and the fingers indicating the direction of the magnetic field. The field generated by a wire coil is depicted in Figure 1.1b. Consider an iron core encased in N wire turns carrying current i, as illustrated in Figure 1.2. The coil's magnetic field will follow the path of least resistance, which is through the iron, similar to how electric current remains within a copper conductor. What Faraday discovered was that the

DOI: 10.1201/ 9781003795841-1

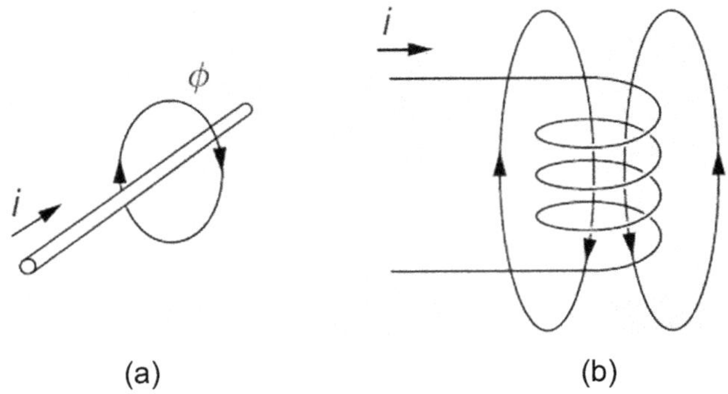

FIGURE 1.1 A magnetic field is formed around a conductor carrying current.

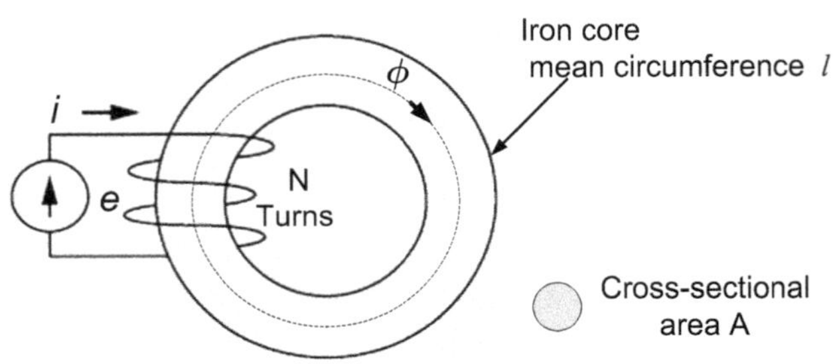

FIGURE 1.2 Basic magnetic circuit.

current passing through the coil not only generates a magnetic field in the iron but also produces a voltage across the coil. This voltage, proportional to the change rate of magnetic flux ϕ in the iron, is referred to as electromotive force, or EMF, denoted by the symbol e.

Assuming that all of the magnetic flux ϕ links all of the turns of the coil, we can write the following important relationship, which is known as *Faraday's law of electromagnetic induction*:

$$e = -N\frac{d\phi}{dt} \tag{1.1}$$

The sign of the induced EMF is always in a direction that opposes the current that created it, a phenomenon referred to as *Lenz's law.*

1.2.2 Simple Magnetic Circuit

Consider a simple structure consisting of a current carrying coil of N turns and a magnetic core of mean length l_c and a cross sectional area A_c as shown in Figure 1.3.

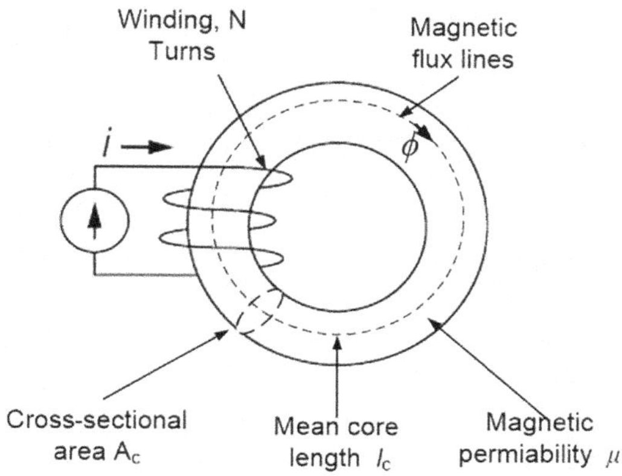

FIGURE 1.3 A simple magnetic circuit.

The permeability of the core material is μ_c. Assume that the size of the device and the operation frequency are such that the displacement current in Maxwell's equations are negligible, and that the permeability of the core material is very high so that all magnetic flux will be confined within the core. By Ampere's law,

$$\oint_c H \bullet dl = \oint_s J \bullet da$$

we can write $H_c l_c = Ni$

where H_c is the magnetic field strength in the core, and Ni the *magneto-motive force*.

The magnetic flux through the cross section of the core can be expressed as

$$\phi_c = B_c A_c \tag{1.2}$$

where ϕ_c is the flux in the core and B_c the flux density in the core.

The constitutive equation of the core material is

$$B_c = \mu H_c \tag{1.3}$$

Therefore, we obtain

$$\phi_c = \frac{Ni}{l_c / (\mu_c A_c)} = \frac{F}{R_c} \tag{1.4}$$

If we take the magnetic flux ϕ_c as the "current," the magneto-motive force F = Ni as the "EMF of a voltage source," and $R_c = l_c / (\mu_c A_c) R_c = l_c /$ (known as the *magnetic reluctance*) as the "resistance" in the magnetic circuit, we have an analog of *Ohm's law* in electrical circuit theory.

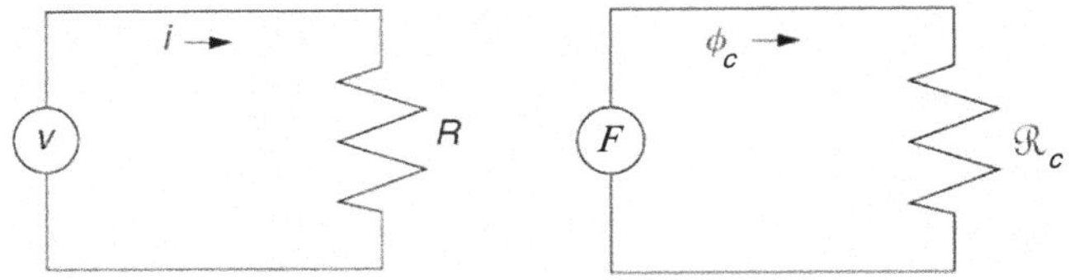

FIGURE 1.4 Equivalent circuits for the electrical and magnetic circuits shown.

1.2.3 Magnetic Circuital Laws

Consider the magnetic circuit in the last section with an air gap of length l_g cut in the middle of a leg as shown in Figure 1.5a. As they cross the air gap, the magnetic flux lines bulge outward somewhat as illustrated in Figure 1.5b. The effect of the *fringing* field is to increase the effective cross sectional area A_g of the air gap. By Ampere's law, we can write

$$F = Ni = H_c l_c + H_g l_g \tag{1.5}$$

where $H_c l_c = \frac{B_c}{\mu_c} l_c = \frac{\phi_c}{\mu_c A_c} l_c = \phi_c R_c$

and

$$H_g l_g = \frac{B_g}{\mu_g} l_g = \frac{\phi_g}{\mu_g A_g} l_g = \phi_g R_g$$

According to Gauss' law in magnetics,

$$\oint_s B \bullet da = 0 \tag{1.6}$$

we know $\phi = \phi_c = \phi_g$

Therefore, $F = \left(R_c + R_g\right)\phi$

That is, the magnetic circuit with an air gap is analogous to a series electric circuit shown in Figure 1.6.

Further, if we regard $H_c l_c$ and $H_g l_g$ as the "voltage drops" across the reluctance of the core and airgap respectively, the above equation from Ampere's law can be interpreted as an analog to the *Kirchhoff's voltage law* (KVL) in electric circuit theory, or

$$\sum F_k = \sum R_k \phi_k \tag{1.7}$$

Around a closed path in a magnetic circuit, the algebraic sum of ampere-turns is equal to the algebraic sum of the products of the reluctances and fluxes.

The *Kirchhoff's current law* (KCL) can be derived from the Gauss' law in magnetics.

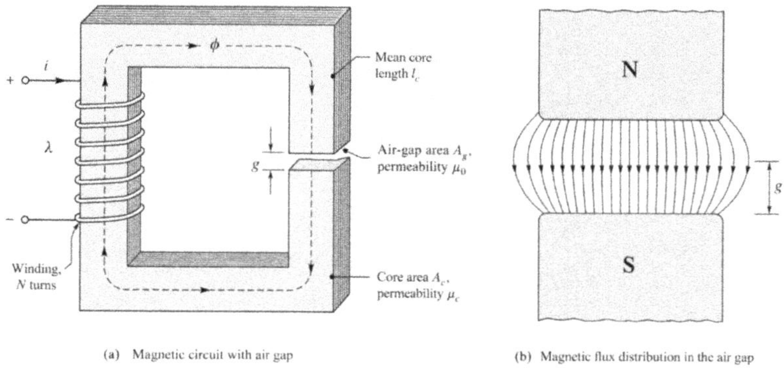

FIGURE 1.5 A magnetic circuit with an air gap.

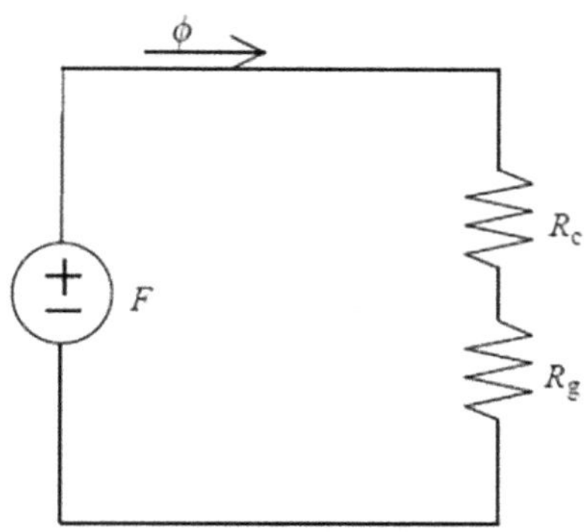

FIGURE 1.6 Series magnetic circuit.

When the Gauss' law is applied to a junction, we have

$$\sum \phi_k = 0 \tag{1.8}$$

which states that *the algebraic sum of all the magnetic fluxes flowing out of a junction in a magnetic circuit is zero.*

Example 1.1: A rectangular shaped iron core has an air-gap of 0.01 *cm*. The mean length path through iron is 39.99 *cm*. The relative permeability of iron is 2,000. The coil has 1,000 turns. The cross-sectional area of the core is 9 cm^2. Calculate the current required to produce a flux of 1 *mWb* in the core.

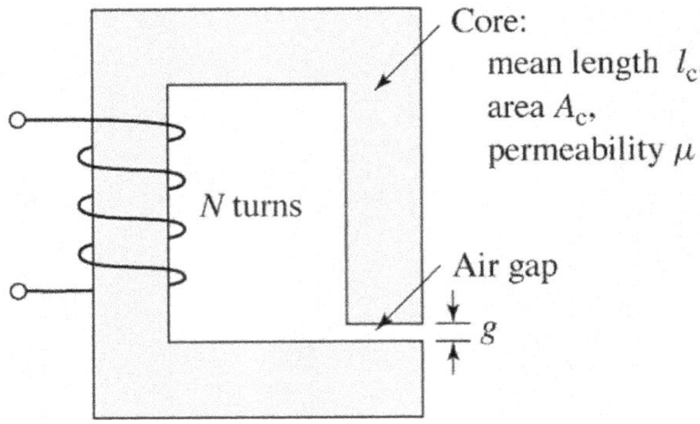

SOLUTION:

The reluctance of core is:

$$R_c = \frac{l_c}{\mu_c A_c} = \frac{0.3999}{2000 * 4\pi * 10^{-7} * 0.0009} = 176794.66 At / Web$$

The reluctance of air-gap is:

$$R_g = \frac{l_g}{\mu_0 A_g} = \frac{0.0001}{4\pi * 10^{-7} * 0.0009} = 88419.14 At / Web$$

Total reluctance:

$$R_t = R_c + R_g = 265214 At / Web$$

The total ampere-turn is:

$$AT = \varnothing R_t = 0.001 \times 265214 = 265.214 At$$

The exciting current:

$$I = \frac{AT}{N} = \frac{265.214}{1000} = 0.265214 A$$

Example 1.2: A circular ring of magnetic material has a mean length of 1.0 m and a cross-sectional area of 0.001 m^2. A saw cut of 5 mm width is made in the ring. Calculate the current required to produce a flux of 1 mWb in the air gap if the ring is wound uniformly with a coil of 200 turns. Take the relative permeability of the ring material $\mu_r = 500$ and neglect leakage and fringing.

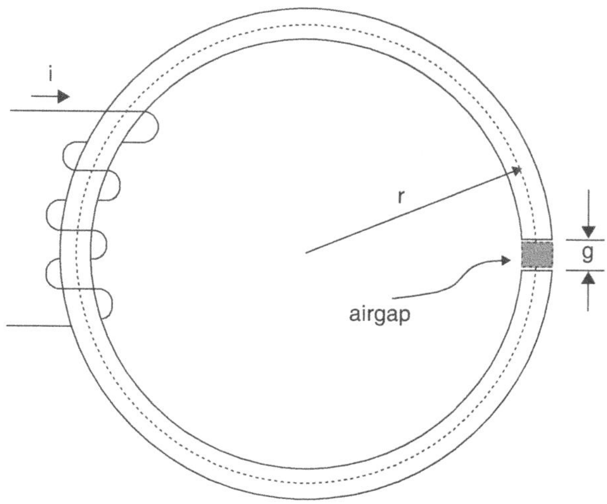

SOLUTION:

The reluctance of the core is:

$$R_c = \frac{l_c}{\mu_c A_c} = \frac{1}{500 \times 4\pi \times 10^{-7} \times 0.001}$$

$$= 1591549.43\, At / Web$$

The reluctance of air-gap is:

$$R_g = \frac{l_c}{\mu_0 A_g} = \frac{1}{4\pi \times 10^{-7} \times 0.001}$$

$$= 397887.36\, At / Web$$

Total reluctance:

$$R_t = R_c + R_g = 1989436.79 At / Web$$

The ampere-turn for core is:

$$AT_c = \varnothing_c R_c = 0.001 \times 1591549.43 = 1591.549\, At$$

The ampere-turn for gap is:

$$AT_g = \varnothing_g R_g = 0.001 \times 397887.36 = 3978.9\, At$$

The total ampere-turn is:

$$AT = 1591.549 + 397.89 = 5569At$$

The exciting current:

$$I = \frac{AT}{N} = \frac{5569}{200} = 27.35A$$

1.3 MAGNETIC MATERIALS

As shown in the previous section, the magnetic flux density $B\left(\text{Wb/m}^2\right)$ and the magnetic field intensity H (A/m) are related through the relationship

$$B = \mu H \tag{1.9}$$

The free space permeability μ_o is a constant given by $4\pi \times 10^{-7}$ H/m. The same value holds good for air and any nonmagnetic materials. For a linear magnetic material which exhibits a straight-line relationship between B and H and the permeability is a constant given by the slope of the B-H curve. However, the B-H curve is nonlinear in general, and the variation of B with H is depicted by the saturation curve of Figure 1.7. The slope of the curve clearly depends on the operating flux density. The factor B/H is not constant, and it is different at various points on the slope because the slope varies. Therefore, the permeability can be defined only at given points on the curve. The different slopes are indicated by tangents drawn on the curve shown in Figure 1.7.

Two examples of slope measurement are shown by the small triangles that are drawn at two points on the slope given in Figure 1.8. Here, a small change in B for a small change in H is being measured; that is to say, the length of the vertical line of triangle is plotted against the length of its base. The smaller the triangle, the more accurate is the measurement. This can be done by using calculus, which deals is in infinitely small changes.

1.3.1 Ferromagnetic Materials

All ferromagnetic materials exhibit a phenomenon called saturation in which the flux density increases in proportion to the field intensity until it cannot do so any longer. Some actual B-H curves for common ferromagnetic materials are shown in Figure 1.9. The basic idea behind magnetic materials is that the spin of electrons constitutes motion charge, and therefore leads to magnetic effects. In most materials, the electron spins cancel out on the whole and no net effect remains. In ferromagnetic materials, however atoms can align so that the electron spins cause a net magnetic effect.

All ferromagnetic materials have atomic magnetic moments that are aligned parallel to each other within small regions called domains. Within these domains, the

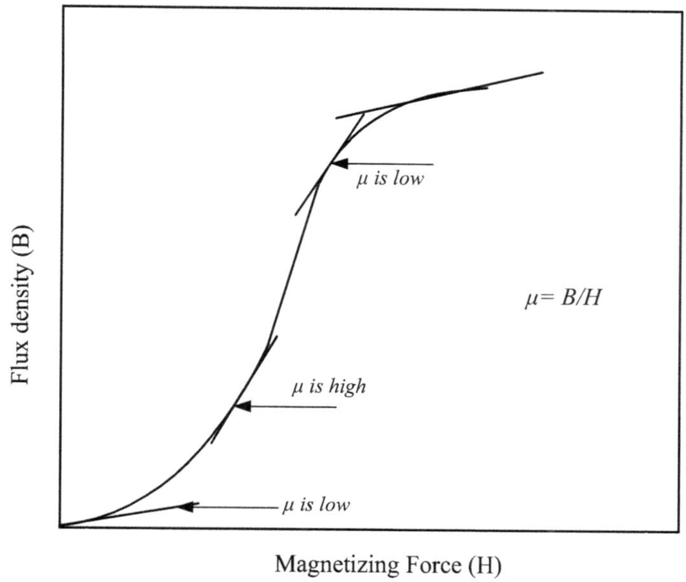

FIGURE 1.7 Magnetization curve of magnetic material.

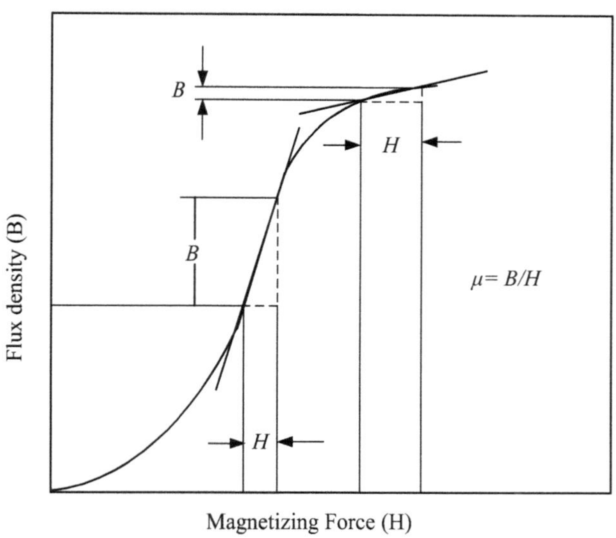

FIGURE 1.8 ***B-H*** Curve.

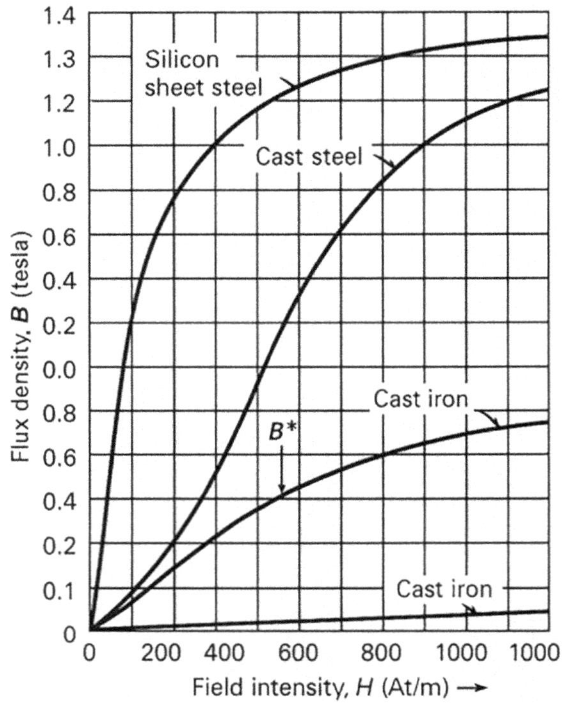

FIGURE 1.9 *B-H* curves for common ferromagnetic materials.

spontaneous magnetization present is equal to saturation magnetization of the material, and so the individual domains are fully magnetized at all times. In the absence of an applied field, there is no net magnetic moment or field generated by the material because the magnetization direction of each domain is randomly oriented.

During magnetization of the material, domains whose magnetization directions have a component in the direction of applied field will grow at the expense of those that do not. As the field increases, more and more domains become aligned. When all of these domains have become aligned, any further increase in magnetic field intensity does not yield an increase in flux density beyond the increase that would be caused in nonmagnetic material. Figure 1.10 indicates the process of magnetization.

1.3.2 Sinusoidal Excitation

It is the same circuit, shown in Figure 1.2, that is excited from an alternating current source, where the magnetic field intensity H will be a function of time (H is directly proportional to the exciting current $i(t)$). The B-H curve obtained in this case is called hysteresis loop as shown in Figure 1.11.

The curve in Figure 1.11 shows that when H is zero, the flux density has certain magnitude which will be acting in the core. This value of flux density which remains

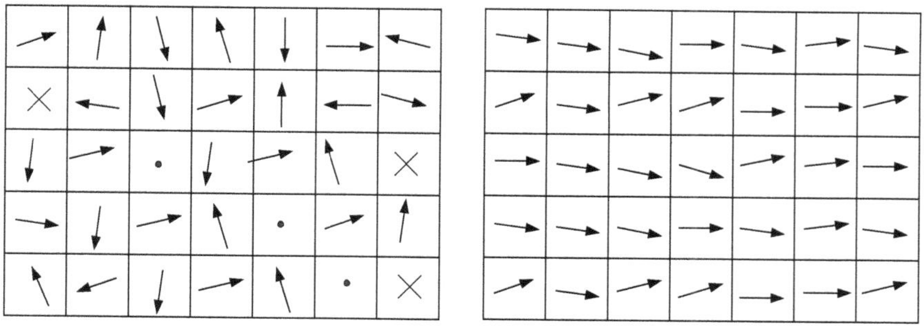

FIGURE 1.10 Magnetization process.

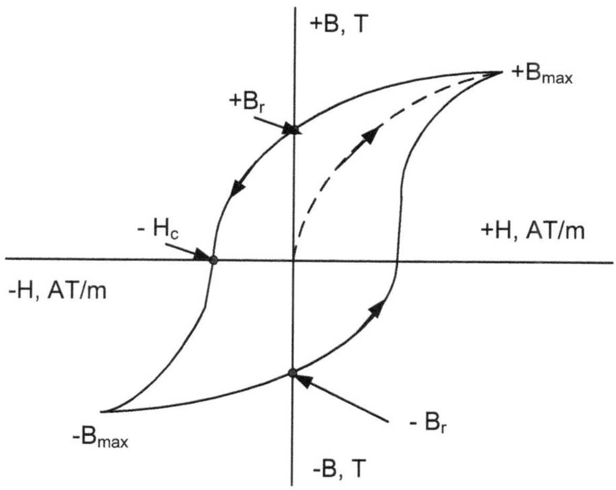

FIGURE 1.11 The hysteresis loop.

in the iron loss after the field intensity has ceased is usually called residual magnetism. It is this residual flux density that makes it possible to develop the large variety of electrical machines.

1.3.3 The Leakage Flux

In any magnetic circuit, there will be two types of magnetic flux. One which is following the desired path and is known as the useful flux, while the other, which is following the undesired path, is called the leakage flux. It must be realized that useful flux is responsible for developing the magnetic bridge inside any electrical machine. This bridge is a necessary element in the process of energy conversion. Thus, an electric machine is designed for an MMF capable of developing both the useful flux and the leakage flux.

1.3.4 The Iron Losses

When magnetic circuit is subjected to time varying flux, certain amount of power loss will be produced in the ferromagnetic material. These power losses are known as core losses, which consists of hysteresis and eddy current losses. These losses are significant in determining the heating, rating, and efficiency of electrical machines, power transformer, and AC-operated magnetic devices. It must be realized that no such losses occur in iron cores that carry magnetic flux, which does not vary with time. Higher values of core losses will tend to raise the temperature of the iron core, up to limits which can ruin the magnetizing prosperities of this core.

The first loss is associated with the phenomenon of hysteresis. If the magnetizing force applied to a magnetic core is changed, through a complete cycle, from $+H_{max}$ to $-H_{max}$ and back to $+H_{max}$, the *B-H* characteristics is described by a hysteresis loop shown in Figure 1.11. The area of the loop represents the energy loss in a unit volume of the core material during one cycle. Thus, the hysteresis loss can be presented by:

$$P_h = (\text{area of the hysteresis loop}) \times (\text{supply frequency}) \times (\text{volume of iron core})$$

$$P_h = K_h \times (\text{volume of iron core}) \times f \times B_m^n \quad \text{W} \tag{1.10}$$

where

K_h is a characteristics constant of the core,
B_m is the maximum flux density in the core,
n is called Steinmetz exponent which may be vary between 1.5 and 2.5 for different materials.

The second loss arises from the fact that the core itself is composed of conducting material, so that the voltages induced in it by the changing flux produce circulating currents in the iron. These circulating currents are called eddy currents and accompanied by loss in the core, called eddy-current loss. However, the iron in the magnetic circuit is laminated to prevent excessive eddy currents where the plane of the laminations is parallel to the flux. Thus, confining the eddy currents to paths of small cross section and correspondingly high resistance. Eddy-current loss is given by the expression

$$P_e = K_e f^2 B_m^2 t^2 \quad \text{W} \tag{1.11}$$

where

K_e is a characteristics constant of the core,
t is the lamination thickness.

The sum of the hysteresis and eddy current losses is equal to the core loss.

$$P_c = P_e + P_h \quad \text{W} \tag{1.12}$$

Example 1.3: A cylinder of iron of volume 8×10^{-3}m revolves for 20 minutes at a speed of 3,000 rpm in a 2-pole field of flux density 0.8 Wb/m^2. If the hysteresis coefficient of iron is 753.6 joule/m^3, specific heat of iron is 0.11, the loss due to eddy current is equal to that due to hysteresis and 25% of the heat produced is lost by radiation, find the temperature rise of iron. Take density of iron as 7.8×10^3 kg/m.

SOLUTION:

An armature revolving in a multipolar field undergoes one magnetic reversal after passing under a pair of poles. In other words, number of magnetic reversals in the same as the number of pair of poles. If P is the number of poles, the magnetic reversals in one revolution are P/2. If speed of armature rotation is N rpm, then number of revolutions/second = N / 60.

$$\text{Revolutions/seconds} = \frac{P}{2}\frac{N}{60} = \frac{P \times N}{120}$$

Here

$$N = 3{,}000\ rpm; P = 2$$

$$\therefore f = \frac{3000 \times 2}{120} = 50 \text{ rps}$$

According to Steinmetz's hysteresis law,

$$W_h = K_h B_{\max}^{1.6} fV \quad \text{W}$$

Note that f here stands for magnetic reversals/second and not for mechanical frequency armature rotation.

$$211 = W_h = 753{,}6 * (0.8)^{1.6} * 50 \times 8(10)^{-3}$$

$$\text{Loss in 20 minutes} = 211 \times 1{,}200 = 253.2 \times 10^3 \text{ J}$$

$$\text{Eddy current loss} = 253.2 \times 10^3 \text{ J};$$

$$\text{Total loss} = 506.4 \times 10^3 \text{ J}$$

$$\text{Heat produced} = 506.4 \times 10^3 / 4200 = 120.57 \text{ kcal};$$

$$\text{Heat utilized} = 120.57 \times 0.75 = 90.43 \text{ kcal}$$

$$\text{Heat absorbed by iron} = \left(8 \times 10^{-3} \times 7.8 \times 10^3\right) \times 0.11 \times t \quad \text{kcal}$$

$$\therefore \text{Temp.} = 13.17\ °\text{C}$$

PROBLEMS

1.1 The maximum working flux density of a lifting electromagnet is 1.8 T and the effective area of a pole face is circular in cross-section. If the total magnetic flux produced is 353 mWb, determine the radius of the pole.

1.2 Determine the magnetic field strength and the MMF required to produce a flux density of 0.25 T in an air gap of length 12 mm.

1.3 A coil of 300 turns is wound uniformly on a ring of non-magnetic material. The ring has a mean circumference of 40 cm and a uniform cross-sectional area of 4 cm^2. If the current in the coil is 5 A, calculate (a) the magnetic field strength, (b) the flux density and (c) the total magnetic flux in the ring.

1.4 A mild steel ring has a radius of 50 mm and a cross-sectional area of 400mm^2. A current of 0.5 A flows in a coil wound uniformly around the ring and the flux produced is 0.1 mWb. If the relative permeability at this value of current is 200, find
(a) the reluctance of the mild steel
(b) the number of turns on the coil

1.5 At what velocity must a conductor 75 mm long cut a magnetic flux of density 0.6T if an EMF of 9V is to be included in it? Assume the conduct or, the field and the direction of motion are mutually perpendicular.

1.6 A conductor moves with a velocity of 15 m/s at an angle (a) 90°; (b) 60°; and (c) 30° to a magnetic field produced between two square-faced poles of side length 2 cm. If the flux leaving a pole face is 5μWb, find the magnitude of the induced EMF in each case.

1.7 A flux density of 1.2 Wb/m^2 is required in the 2 mm air gap of an electromagnet having an iron path 1 m long. Calculate the MMF required, assuming a relative permeability of iron as 1,500. Neglect leakage.

1.8 Find the speed that a conductor of length 120 mm must be moved at right angles to a magnetic field of flux density 0.6 T to induce in it an EMF of 1.8 V.

1.9 A 25 cm long conductor moves at a uniform speed of 8 m/s through a uniform magnetic field of flux density 1.2 T. Determine the current flowing in the conductor when (i) its ends are open-circuited (ii) its ends are connected to a load of 15 ohm resistance.

1.10 A conductor of length 0.5 m situated in and at right angles to a uniform magnetic field of flux density 1 Wb/m^2 moves with a velocity of 40 m/s. Calculate the EMF induced in the conductor. What will be the EMF induced if the conductor moves at an angle 60° to the field.

1.11 A circular iron ring has a mean circumference of 1.5 m and cross-sectional area of 0.01 m^2. A saw cut of 4 mm wide is made in the ring. Calculate the magnetizing current required to produce a flux of 0.8 mWb in the gap if the ring is wound with a coil of 175 turns. Assume relative permeability of iron as 400 and leakage factor 1.25.

1.12 An iron ring has a mean diameter of 15 cm, a cross-sectional area of 20 cm^2 and a radial gap of 0.5 mm cut in it. It is uniformly wound with 1,500 turns of insulated wire and a magnetizing current of 1 A produces a flux of 1 mWb. Neglecting the effect of magnetic leakage and fringing, calculate:
(i) Reluctance of the magnetic circuit.
(ii) Relative permeability of iron.

2 Fundamentals of Electrical Machines

2.1 ELECTROMECHANICAL FUNDAMENTALS

Energy conversion means converting one form of energy into another form. An electric generator converts mechanical energy (drawn from prime mover through shaft) into electric energy. An electric motor converts electric energy into mechanical energy (which drives mechanical load, e.g., fan, lathe, etc.).

Electric generators and motors operate by virtue of induced EMF. The induction of EMF is based on Faraday's law of electromagnetic induction. Every generator and motor has a stator (which remains stationary) and rotor (which rotates).

2.1.1 Faraday's Law

The electric fields and magnetic fields considered up to now have been produced by stationary charges and moving charges (currents), respectively. Imposing an electric field on a conductor gives rise to a current which in turn generates a magnetic field. One could then inquire whether or not an electric field could be produced by a magnetic field. In 1831, Michael Faraday discovered that, by varying magnetic field with time, an electric field could be generated. The phenomenon is known as electromagnetic induction. Figure 2.1 illustrates one of Faraday's experiments.

Faraday showed that no current is registered in the galvanometer when bar magnet is stationary with respect to the loop. However, a current is induced in the loop when a relative motion exists between the bar magnet and the loop. In particular, the galvanometer deflects in one direction as the magnet approaches the loop, and the opposite direction as it moves away. Faraday's experiment demonstrates that an electric current is induced in the loop by changing the magnetic field. The coil behaves as if it were connected to an EMF source. Experimentally, it is found that the induced EMF depends on the rate of change of magnetic flux through the coil as follow:

$$e = N\frac{d\phi}{dt} \tag{2.1}$$

where

e = EMF induced, Vs
N = number of turns
ϕ = flux, weber
t = time, seconds.

DOI: 10.1201/ 9781003795841-2

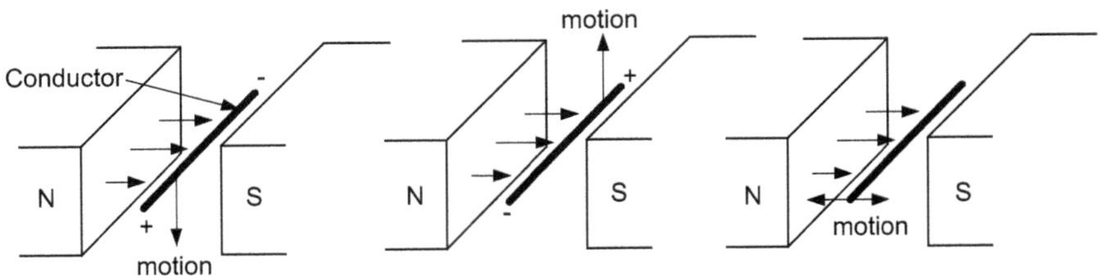

FIGURE 2.1 Magnetic induction.

Example 2.1: The field coils of a six-pole DC generator each having 500 turns are connected in series. When the field is excited, there is a magnetic flux of 0.02 Wb/pole. If the field circuit is opened in 0.02 second and residual magnetism is 0.002 Wb/pole, calculate the average voltage which is induced across the field terminals, in which direction is this voltage directed relative to the direction of the current.

SOLUTION:

$$\text{Total number of turns, N} = 6 \times 500 = 3000$$

$$\text{Total initial flux} = 6 \times 0.02 = 0.12 \text{ Wb}$$

$$\text{Total residual flux} = 6 \times 0.002 = 0.012 \text{ Wb}$$

$$\text{Change in flux, } d\phi = 0.12 - 0.012 = 0.108$$

$$\text{Time of opening the circuit, } dt = 0.02 \text{ second}$$

$$\therefore \text{Induced EMF} = N\frac{d\phi}{dt} = 3000 \times 0.108 / 0.02 = 16{,}200 \text{ V}$$

Example 2.2: A coil of resistance 100 Ω is placed in a magnetic field of 1 mWb. The coil has 100 turns and a galvanometer of 400 Ω resistance is connected in series with it. Find the average EMF and the current if the coil is moved in 1/10th second from the given field to a field of 0.2 mWb.

SOLUTION:

$$\text{Induced EMF} = N\frac{d\phi}{dt}$$

$$d\phi = 1 - 0.2 = 0.8 \text{ mWb}$$

$$dt = 1/10 = 0.1 \text{ second; N} = 100$$

$$e = 100 \times 0.8 \times 10^{-3} / 0.1 = 0.8 \text{ V}$$

$$\text{Total circuit resistance} = 100 + 400 = 500 \ \Omega$$

$$\therefore \text{Current induced} = 0.8 / 500 = 1.6 \times 10^{-3} \text{ A} = 1.6 \text{ mA}$$

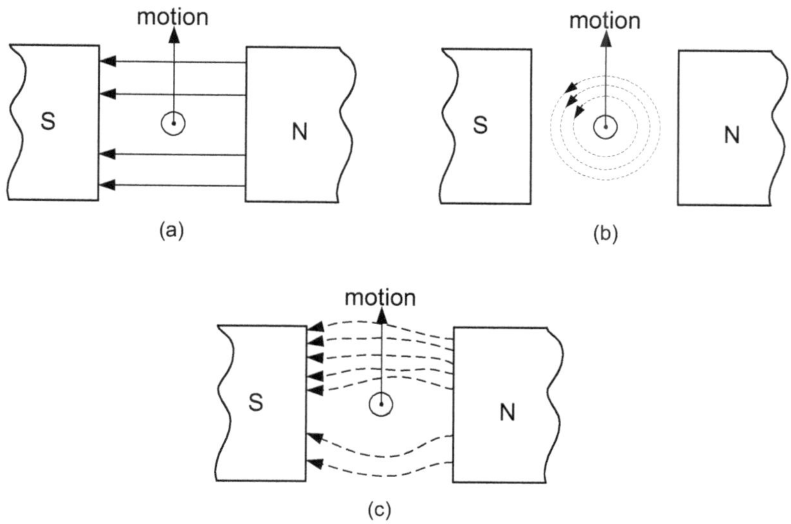

FIGURE 2.2 Illustration of Lenz's Law: (a) Direction of EMF, (b) Field produced by induced current, (c) Resultant magnetic field.

2.1.2 Lenz's Law

The directions of induced voltage and current flowing in a conductor moving in a magnetic field can be verified by Lenz's Law. Lenz's law stated that: In all cases of electromagnetic induction, an induced voltage will cause a current to flow in a closed circuit in such a direction that its magnetic effect will oppose the change that produced it.

The above statement of Lenz's law implies:

(1) a cause (a change in flux linkage)
(2) an effect of opposing any change of current itself.

So, in all cases of electromagnetic induction, wherever a change in flux linkage occurs, a voltage is induced that tend to set up a current in such a direction as to produce a field which opposes the change in flux linking the turns of the circuit.

In fact, when a circuit or component possesses the property of opposing any change of current in itself, that property is called inductance, and the induced EMF is termed as an EMF of self-induction. As an illustration, consider the conductor shown in Figure 2.2 which is being moved by some prime mover in the upward direction as shown. If some electric load is connected to this elementary generator, current will tend to flow in the conductor, in the same direction as the EMF, producing a magnetic field around the conductor as shown. The resultant magnetic field will tend to oppose the upward motion of the conductor.

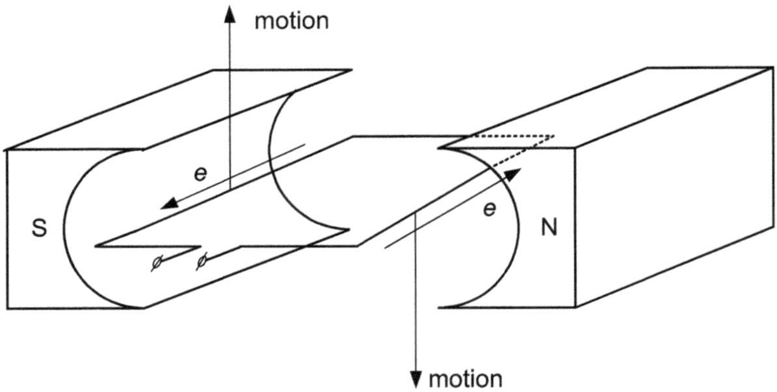

FIGURE 2.3 Simple AC generator.

2.1.3 Basic AC Generator

An AC generator uses the principle of Faraday's electromagnetic induction to convert mechanical energy, such as rotation, into electrical energy, a sinusoidal waveform. A simple generator consists of a pair of permanent magnets producing a fixed magnetic field between a north and a south pole. Inside this magnetic field is a single rectangular loop of wire that can be rotated around a fixed axis, allowing it to cut the magnetic flux at various angles as shown in Figure 2.3.

As the coil rotates anticlockwise around the central axis, which is perpendicular to the magnetic field, the wire loop cuts the lines of force set up between the north and south poles at different angles as the loop rotates. The amount of induced EMF in the loop at any instant of time is proportional to the angle of rotation of the wire loop. As the loop rotates, electrons in the wire loop flow in one direction around the loop. When the wire loop moves across the magnetic lines of force in the opposite direction, the electrons in the wire loop flow in the opposite direction. Then the direction of the electron movement determines the polarity of the induced voltage. When the loop or coil rotates one complete revolution, or 360°, one full sinusoidal waveform is produced with one cycle of the waveform being produced for each revolution of the coil. As the coil rotates within the magnetic field, the electrical connections are made to the coil by means of carbon brushes and slip-rings which are used to transfer the electrical current induced in the coil.

The amount of EMF induced into a coil cutting the magnetic lines of force is determined by the following three factors:

- Speed – the speed at which the coil rotates inside the magnetic field.
- Strength – the strength of the magnetic field.
- Length – the length of the coil or conductor passing through the magnetic field.

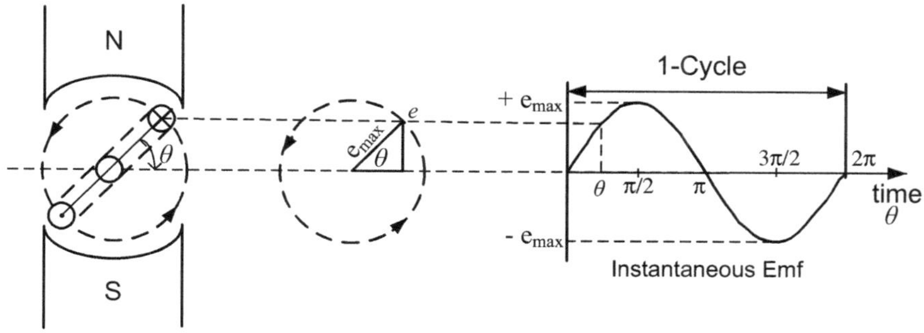

FIGURE 2.4 Armature coil revolving in 2-pole magnetic field.

We know that the frequency of a supply is the number of times a cycle appears in one second and that frequency is measured in Hertz. As one cycle of induced EMF is produced each full revolution of the coil through a magnetic field comprising of north and south poles as shown above, if the coil rotates at a constant speed, a constant number of cycles will be produced per second giving a constant frequency. So by increasing the speed of rotation of the coil, the frequency will also be increased. Therefore, frequency is proportional to the speed of rotation, $(f \propto \mathrm{N})$, where N = *rpm*.

Also, our simple single coil generator above only has two poles, one north pole and one south pole, giving just one pair of poles. If we add more magnetic poles to the generator above so that it now has four poles in total, two north and two south, then for each revolution of the coil two cycles will be produced for the same rotational speed. Therefore, frequency is proportional to the number of pairs of magnetic poles $(f \propto P)$ of the generator where $P =$ is the number of "pairs of poles."

2.1.4 Rotating Coil in a Magnetic Field

The EMF induced in the coil at any instant of time depends upon the rate or speed at which the coil cuts the lines of magnetic flux between the poles, and this is dependent upon the angle of rotation, Theta (θ), of the generating device. Because an AC waveform is constantly changing its value or amplitude, the waveform at any instant in time will have a different value from its next instant in time. For example, the value at 1 ms will be different to the value at 1.2 ms and so on. These values are known generally as the Instantaneous Values, or V_i Then the instantaneous value of the waveform and also its direction will vary according to the position of the coil within the magnetic field as shown in Figure 2.4.

The instantaneous value of a sinusoidal waveform is given as the and this is generalized by the formula.

$$E = e_{max} sin\theta \tag{2.2}$$

where, e_{max} is the maximum voltage induced in the coil and $\theta = \omega t$, is the angle of coil rotation.

If the maximum or peak value of the waveform is known, by using above the formula the instantaneous values at various points along the waveform can be calculated. A sinusoidal waveform shape can be constructed. In order to keep things simple, we will plot the instantaneous values for the sinusoidal waveform at every 45° and assume a maximum value of e_{max}. Plotting the instantaneous values at shorter intervals, for example, at every 30°, would result in a more accurate waveform construction.

Refer to Figure 2.4 The EMF induced in the conductor is given by:

$$E = Blv\ V \tag{2.3}$$

where $B =$ flux density of the magnitude field in tesla (T),
$l =$ length of the conductor in meters (m), and
$v =$ velocity of the conductor in m/s

from equation (2.2), the induced EMF,

$$E = Blv\ sin\,(\theta)\ V \tag{2.4}$$

The direction of the EMF is given by Fleming's right hand rule.

2.1.5 Relation Between n_s, P, and F

The frequency of the generated EMF depends on the number of poles on the rotor and the speed at which the rotor is driven.

In the simple 2-pole alternator of Figure 2.5, one complete cycle of EMF is generated when a pair of rotor poles (a north and a south pole) is moved past a coil. This occurs in one revolution of the rotor.

In practice, however, alternators may have several pairs of poles. The actual relationship between the number of cycles generated in one revolution of the rotor and the time in seconds is given in the following formula:

$$N = 60f\ /\ P \tag{2.5}$$

where $f =$ frequency in hertz (cycles per second) and
$N =$ rotor speed in revolutions per minute.

For the standard frequency of 50 Hz, the highest possible speed for a 2-pole machine where $P = 1$ (Figure 2.5) is 50 rps or 3,000 rpm. High-speed turbo-alternators with outputs up to 50,000 VA are built as 2-pole machines (3,000 rpm), but for higher outputs, 4-pole alternators running at 1,500 rpm are more common. Slow-speed alternators, on the other hand, have many poles (up to 48).

A four-pole rotor, however, has two north and two south poles Figure 2.6. The stator of a four-pole alternator has two coils per phase connected together to form a

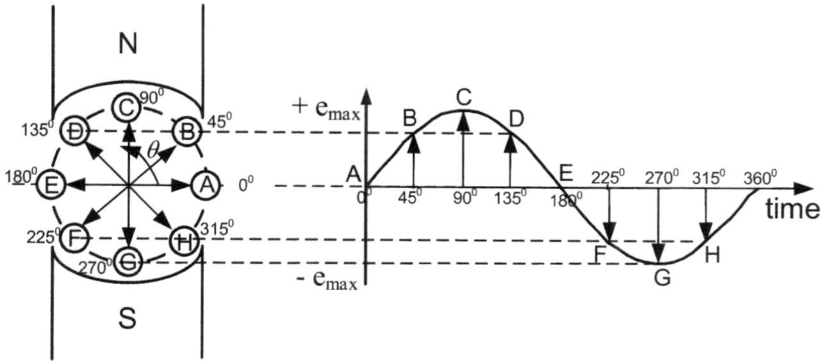

FIGURE 2.5 Armature coil revolving in 4-pole magnetic field.

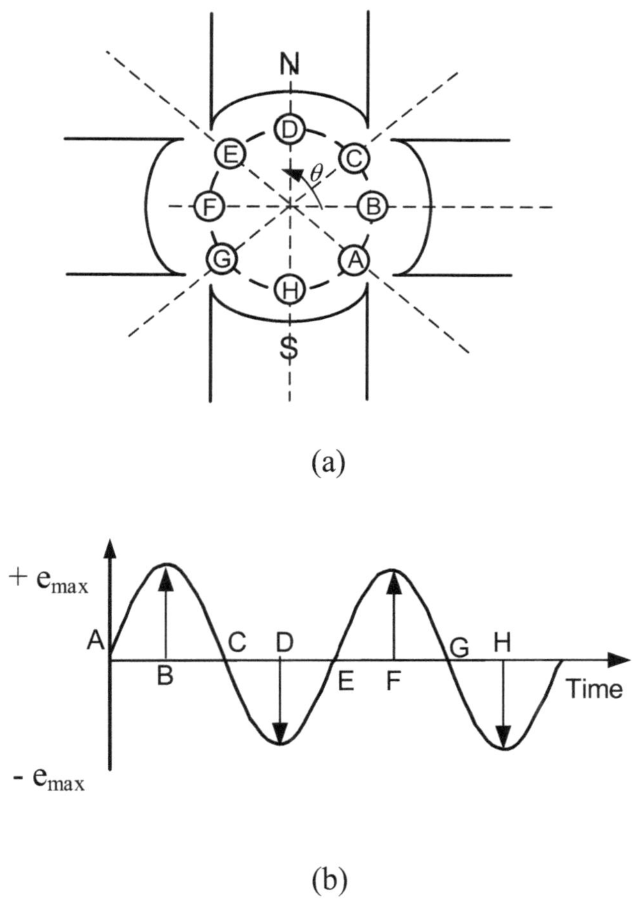

FIGURE 2.6 Generated EMF in a single coil.

chain winding, as shown in Figure 2.6. For every revolution of the rotor (360 mechanical degrees), two cycles of EMF (720 electrical degrees) are generated in each phase compared with one cycle in the two-pole alternator.

The mechanical angle θ_m round the machine periphery is always 2π but it is more convenient to designate the complete angle of one cycle of the B-wave as 2π, which to distinguish it is called electrical angle θ. For a P-pole, the relationship between these two angles is expressed as

$$\theta / \theta m = 2\,\pi \times P / 2\,\pi = P \tag{2.6}$$

2.2 ELECTROMAGNETIC FORCE ON A CONDUCTOR

If a conductor is placed in a magnetic field and current is allowed to flow through the conductor, it will tend to move. This principle is sometimes called "motor action." The previous example indicates that the force acting on the conductor depends upon three factors. First a magnetic field must exist (which is presented by the flux density B), then a conductor which is presented by its active length l must be located in the magnetic field, and finally a current must flow in the conductor. The direction of the electromagnetic force developed is determined by applying Fleming's left hand rule. For this rule, the thumb, index finger, and the middle finger are held at right angle at each other. The index finger is used to indicate the direction of flux. The middle finger points in the direction of the current, and the thumb points in the direction of force on the conductor.

The electromagnetic force acting on a current carrying conductor can be presented by:

$$F_e = BIl \text{ Newton} \tag{2.7}$$

where

$B =$ flux density, tesla (T)
$l =$ active length of conductor, meters (m)
$I =$ current, (A).

2.2.1 Electromagnetic Torque

Torque is defined as that force which tends to produce and maintain rotation. The function of torque in a DC motor is to provide the mechanical output or drive the piece of equipment that the DC motor is attached to. When a voltage is applied to a motor, current will flow through the field winding, establishing a magnetic field. Current will also flow through the armature winding, from the negative brush to the positive brush.

Since the armature is a current-carrying conductor in a magnetic field, the conductor has a force exerted on it, tending to move it at right angles to that field. Using the left-hand rule for current- carrying conductors, you will see that the magnetic field on one side is strengthened at the bottom, while it is weakened on the other side. Using the right-hand rule for motors, we can see that there is a force exerted on the armature which tends to turn the armature in the counter-clockwise direction. The sum of the forces multiplied by the radius of the armature is equal to the torque developed by the motor.

It is evident that if the armature current was reversed, but the field was the same, torque would be developed in the opposite direction. Likewise, if the field polarity was reversed and the armature remained the same, torque would also be developed in the opposite direction. The force that is developed on a conductor of a motor armature is due to the combined action of the magnetic fields. The force developed is directly proportional to the strength of the main field flux and the strength of the field around the armature conductor. As we know, the field strength around each armature conductor depends on the amount of current flowing through the armature conductor.

PROBLEMS

2.1 A coil comprised from 240 turns is wrapped around a ferrite core carrying a peak flux density of 0.2 T varying sinusoidally at a frequency of 12 kHz. The cross sectional area of the core is 36 mm^2. What is the magnitude of:
- (a) The flux in the core.
- (b) The voltage induced in the coil.

2.2 A toroidal ring core is used to create an inductor by having a coil of 400 turns wrapped around it. The toroid has a cross section of 25 cm^2 a circumferential length of 160 mm and a relative permeability of 120. The coil carries a current of 1.2 A.
- (a) What is the MMF of the coil?
- (b) What is the magnitude of the magnetizing field strength in the core?
- (c) What is the magnitude of the magnetic flux density in the core?
- (d) What is the total flux in the core?
- (e) What is the flux-linkage with the coil?
- (f) What is the inductance of the coil?
- (g) What is the reluctance of the core?
- (h) How much magnetic energy is stored in the core?
- (i) If the core saturates at a flux density of 1.5 T, what is the coil current required to produce this?
- (j) What will happen to the inductance if the coil current rises above this value?

2.3 A conductor carries a current of 20 A and is at right-angles to a magnetic field having a flux density of 0.9 T. If the length of the conductor in the field is 30 cm, calculate the force acting on the conductor. Also determine the value of the force if the conductor is inclined at an angle of 30° to the direction of the field.

2.4 Determine the current required in a 400 mm length of conductor of an electric motor, when the conductor is situated at right-angles to a magnetic field of flux density 1.2 T if a force of 1.92 N is to be exerted on the conductor. If the conductor is vertical, the current flows downward, and the direction of the magnetic field is from left to right, what is the direction of the force?

2.5 A conductor 350 mm long carries a current of 10 A and is at right-angles to a magnetic field lying between two circular pole faces each of radius 60 mm. If the total flux between the pole faces is 0.5 mWb, calculate the magnitude of the force exerted on the conductor.

$$\text{"Instantaneous value} = \text{Maximum value} \times \sin\theta\text{"}$$

3 Power Transformer

3.1 INTRODUCTION

The transformer is the device primarily responsible for the wide utilization of alternating current today. Basically, the transformer is a machine for transferring electrical energy from one circuit to another without a change in frequency. The transfer of energy usually takes a place with a change in voltage. The transformer accomplishes this change in voltage without the use of moving parts. The transformer is the most efficient piece of electrical machinery and efficiencies of 98–99% are very common. Another important factor is that, since there are no moving parts, maintenance is simpler and cheaper, and the required insulation is easily constructed. In power systems, connecting a step-up transformer between the alternator and a transmission line permits a practical design voltage for the alternator and at the same an efficient transmission line voltage. Step-down transformers connected between the transmission line and the various electrical loads; permit the transmitted power to be used at a safe voltage. Generally, large power systems existing today could not have been developed without transformers.

3.2 TRANSFORMER PRINCIPLES OF OPERATIONS

A clear understanding of how transformers work is necessary in order to wire them properly in an electrical system. Understanding input and output current and grounding are particularly troublesome. A dual-voltage transformer can be ruined when power is applied if the connections are made improperly. An important property of electricity is that a magnetic field is produced around a current-flowing wire. The more current that flows, the stronger is the magnetic field. An even stronger magnetic field can be produced by winding the wire into a coil. Now the magnetic fields of adjacent wires add together to

form one strong magnetic field. The electrical current flowing in a transformer is alternating current. The current flows first in one direction, stops, then reverses, and flows in the other direction. The magnetic field around the winding is constantly in motion. Notice that the north and south poles of the magnetic field reverse when the flow of current reverses.

Another property of electricity is important to the operation of a transformer. When a magnetic field moves across a wire, a voltage is induced into the wire. If the wire forms a complete circuit, current will flow in the wire. If a second coil of wire is placed in a moving magnetic field, then a voltage will be induced in this second coil (Figure 3.1). This phenomenon is called mutual induction. Alternating current in one winding produces a moving magnetic field that induces a voltage in a second winding. Electrical energy is converted into magnetic field and then converted back into electrical energy in a second winding. The trick is to do this with little or no loss

DOI: 10.1201/9781003795841-3

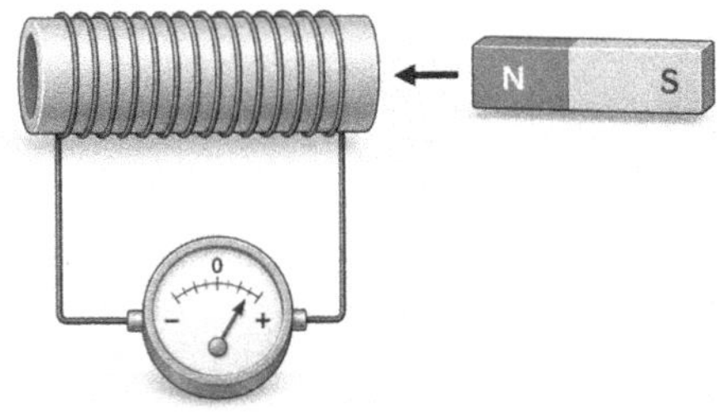

FIGURE 3.1 When the magnetic field moves across this wire, an electrical current flows in the wire.

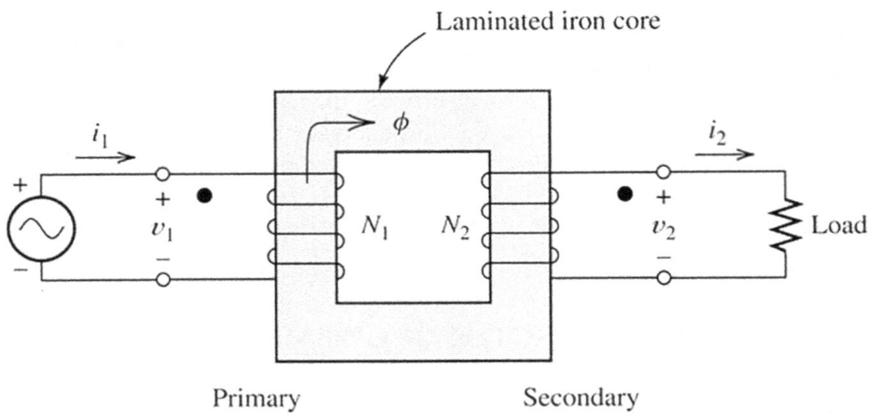

FIGURE 3.2 The two transformer windings are on separate parts of the core.

of energy. The magnetic field loses strength quickly in air therefore; a special steel core is used. The core is com posed of thin sheets of a silicon-steel alloy. The magnetic field is concentrated in the core, and energy losses are reduced to a minimum. Figure 3.2 shows the two windings separated. Most transformers have one winding placed directly over the other to further reduce the loss of energy.

3.3 IDEAL TRANSFORMER

A transformer that possess the following properties are considered to be an ideal transformer.

1. Primary and secondary resistances are assumed to be zero. Hence, there is no power loss and voltage drop in an ideal transformer.
2. Leakage flux is completely absent.
3. The permeability of the core is infinite and so the magnetizing current is zero.
4. The core losses are considered to be zero.

When the secondary winding is open and an alternating voltage V_I is applied to the primary winding, a minimal current known as the no-load current I0 flows, establishing a magnetic flux within the core. This alternating flux interacts with both the primary and secondary coils, inducing electromotive forces E_I and E_2 in them through mutual induction. The formula for the induced EMF E in a coil of N turns is based on the rate of change of flux.

$$E = -N\frac{d\Phi}{dt}$$

where $\frac{d\Phi}{dt}$ is the rate of change of flux.

In an ideal transformer, this rate of change of flux remains consistent for both the primary and secondary, leading to a constant induced EMF per turn. Assuming there are no losses, the voltage and turns ratios, also known as the transformation ratio of the transformer, adhere to the relationship:

$$\frac{E_1}{N_1} = \frac{E_2}{N_2}$$

that is, the induced EMF per turn is constant.

Assuming no losses, $E_1 = V_1$ and $E_2 = V_2$

Hence,

$$\frac{V_1}{N_1} = \frac{V_2}{N_2} \text{ or } \frac{V_1}{V_2} = \frac{N_1}{N_2} \tag{3.1}$$

The voltage ratio $\frac{V_1}{V_2}$ and the turns ratio $\frac{N_1}{N_2}$ are referred to as the transformation ratio $\alpha\alpha$ of the transformer. If N_2 is smaller than $N1$, then V_2 is lower than V_I and the device functions as a step-down transformer. Conversely, if N_2 is greater than $N1$, V_2 exceeds V_I, defining the device as a step-up transformer. When a load is connected to the secondary winding, a current I_2 is generated. Assuming the transformer operates ideally without losses, it is deemed to be 100 % efficient. This means that the input power matches the output power, or $V_I I_I = V_2 I_2$, indicating that in an ideal transformer, the primary and secondary ampere-turns are equal.

Thus

$$\frac{V_1}{V_2} = \frac{I_2}{I_1} = \alpha \tag{3.2}$$

Combining equations (3.1) and (3.2) gives:

$$\frac{V_1}{V_2} = \frac{N_1}{N_2} = \frac{I_2}{I_1} = \alpha \tag{3.3}$$

The capacity of a transformer is expressed in V-amperes, reflecting the amount of power it can handle before overheating. With reference to Figure 3.2, this rating can be represented as V_1I_1 or V_2I_2, where I_2 is the full-load secondary current. If the mutual flux is considered to oscillate sinusoidally with the frequency of the supply voltage, it is described as follows:

$$\Phi = \phi_m \sin 2\pi ft \tag{3.4}$$

The equation for the instantaneous EMF e_1 can be

$$e_1 = N_1 \frac{d(\phi_m \sin 2\pi ft)}{dt} \tag{3.5}$$

$$e_1 = N_1 2\pi f \phi_m \cos 2\pi ft \tag{3.6}$$

The peak value of primary EMF will be

$$E_{p1} = 2\pi f\ N_1 \phi_m \tag{3.7}$$

Thus, the rms value E_1 of the primary will be given by:

$$E_1 = \frac{1}{\sqrt{2}} 2\pi f\ N_1 \phi_m = 4.44 f N_1 \phi_m \tag{3.8}$$

In a similar manner, the rms value E_2 of the primary will be given by:

$$E_2 = 4.44\ f N_2 \phi_m \tag{3.9}$$

Example 3.1: A transformer has 500 primary turns and 3,000 secondary turns. If the primary voltage is 220 V, determine the secondary voltage, assuming an ideal transformer.

SOLUTION:

For an ideal transformer, voltage ratio = turns ratio

that is, $\frac{V_1}{V_2} = \frac{N_1}{N_2}$ hence $\frac{220}{V_2} = \frac{500}{3000}$

Thus, secondary voltage

$$V_2 = 220 \times 3000/500 = 1220\ \text{V}$$

Example 3.2: An ideal transformer with a turns ratio of 2:6 is fed from a 220 V supply. Determine its output voltage.

SOLUTION:

A turns ratio of 2:6 means that the transformer has two turns on the primary for every six turns on the secondary (i.e., a step-up transformer); thus $\frac{N_1}{N_2} = \frac{2}{6}$

For an ideal transformer, $\frac{V_1}{V_2} = \frac{N_1}{N_2}$

Hence $\frac{220}{V_2} = \frac{2}{6}$

Thus, the secondary voltage: $V_2 = 660$ V

Example 3.3: An ideal transformer has a turns ratio of 8:1 and the primary current is 3A when it is supplied at 220V. Calculate the secondary voltage and current.

SOLUTION:

A turns ratio of 8:1 means

$\frac{N_1}{N_2} = \frac{8}{1}$ that is, a step-down transformer.

$\frac{V_1}{V_2} = \frac{8}{1}$ or secondary voltage $V_2 = 220 \times 1/8 = 55$ V

Also, $\frac{N_1}{N_2} = \frac{I_2}{I_1}$ hence secondary current

$I_2 = 3 \times 8/1 = 24$ A

3.4 TRANSFORMER RATINGS

Transformers are rated in volt-ampere (VA) or kilo volt-ampere (kVA). This means that the primary and the secondary winding are designed to withstand the VA or kVA rating stamped on the transformer nameplate. The primary and secondary full-load currents usually are not given. The installer must be able to calculate the primary and secondary currents from the nameplate information. When the volt-ampere (or kilo volt-ampere) rating is given, along with the primary voltage, the primary full-load current can be determined, using Equation (3.10).

$$\text{Full-load current} = \frac{VArating}{Voltage} \tag{3.10}$$

Example 3.4: A single-phase transformer with a 2 kVA rating has a 480 V primary, and a 120 V secondary. Determine the primary and secondary full-load currents of the transformer.

SOLUTION:

Use Equation (3.10) to solve for both primary and secondary currents.

$$\text{Primary full-load current} = \frac{2000}{480} = 4.17\text{A}$$

$$\text{Secondary full-load current} = \frac{2000}{120} = 16.67\text{A}$$

Example 3.5: A 3,810 to 240 V, 50 Hz, 10 kVA single-phase transformer is rated at 2.5 Vs per turn for its winding coils. Assume that it is an ideal transformer and calculate the following:

(a) Step-up transformer ratio
(b) Step-down transformer ratio
(c) Total turns of the high-voltage (HV) winding
(d) Total turns of the low-voltage (LV) winding
(e) Primary current as a step down transformer
(f) Primary current as a up down transformer

SOLUTION:

(a) As a step-up transformer

$$V_1 = 240 \text{ V}, V_2 = 3810 \text{ V}$$

$$\therefore \text{Turns ratio} = \frac{V_1}{V_2} = \frac{240}{3810} = 0.063$$

(b) As a step-down transformer

$$V_1 = 3810 \text{ V}, V_2 = 240 \text{ V}$$

$$\therefore \text{Turns ratio} = \frac{V_1}{V_2} = \frac{3810}{240} = 15.9$$

(c) $\because V_H = (\text{V per turn}).\text{N}_\text{H}$

$\therefore$ high-voltage winding turns

$$N_H = \frac{3810}{2.5} = 1524 \text{ turns}$$

(d) low voltage winding turns

$$N_L = \frac{240}{2.5} = 96\text{turns}$$

(e) As a step-down transformer

$$V_1 = 3810 \text{ V}$$

$$\therefore \text{Primary current } I_1 = \frac{VA}{V_1} = \frac{10\times10^3}{3810} = 2.62\text{A}$$

(f) As a step-down transformer

$V_2 = 240$ V

$$\therefore \text{Secondary current } I_2 = \frac{VA}{V_2} = \frac{10\times10^3}{240} = 41.66 \text{ A}$$

It is sometimes convenient to refer the impedance connected across one side of a transformer to the other side of the transformer. Figure 3.3 shows an impedance Z_L connected across the secondary winding of an ideal transformer.

Now, the secondary terminal voltage is

$$V_2 = I_2 Z_L$$

From eqn. (3.3), it follows that

$$\frac{V_1}{\alpha} = \alpha I_1 Z_L$$

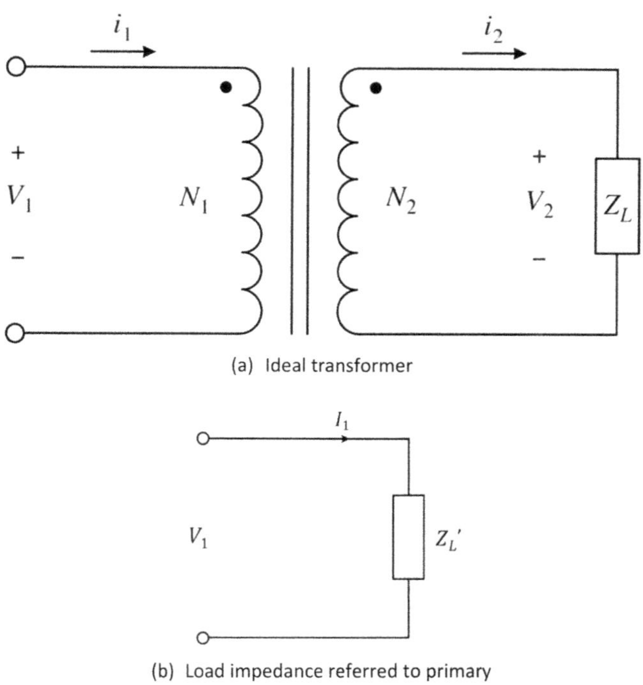

FIGURE 3.3 Equivalent circuit of an ideal transformer.

$$\therefore \frac{V_1}{I_1} = \alpha^2 Z_L$$

But $\frac{V_1}{I_1}$ = the input impedance viewed across the terminals of the primary winding of the transformer.

$$\therefore Z_L' = \alpha^2 Z_L \tag{3.11}$$

3.5 THE PRACTICAL (REAL) TRANSFORMER

Transformer windings are primarily composed of copper, which, despite being a good conductor, has a finite resistance. Both the primary and secondary windings possess distinct resistances, R_1 and R_2 respectively, which are evenly distributed across the windings. These resistances lead to what are known as copper losses, calculated as I^2R. In a practical transformer, besides the mutual flux that connects both primary and secondary windings, there is a flux component that links only one winding and not the other. This type of flux is referred to as leakage flux. The flux that exclusively associates with the primary winding is called primary leakage flux, while the flux that only links with the secondary is known as secondary leakage flux.

Figure 3.4 schematically illustrates both the mutual and leakage flux. Based on our understanding of magnetic circuits, the linkage of flux with a winding leads to the winding's inductance, which is defined as the ratio of flux linkage per ampere. As the flux in a transformer alternates, this linkage induces a voltage in the winding. Consequently, the primary leakage flux, which is proportional to the primary current I_1, results in an induced voltage, manifesting as a voltage drop. The same applies to the secondary leakage flux, as depicted in Figure 3.4. The impacts of these induced electromotive forces are thus characterized as inductive leakage reactances X_1 and X_2.

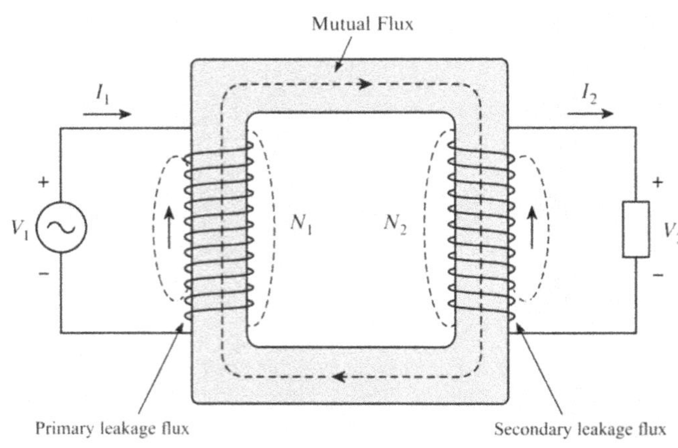

FIGURE 3.4 Mutual and leakage flux paths.

Both the resistance and the leakage reactance of the transformer windings are series effects and at operating frequencies, which is very low (50 Hz/60 Hz), these can be regarded as the lumped parameters. Hence, the transformer consists of lumped resistances R_1 and R_2; and reactance X_1 and X_2 in series with corresponding windings. Because of the presence of these lumped quantities the induced E_1 and E_2 may vary from the secondary voltages V_1 and V_2 because of the small voltage drops in the winding resistances and leakage reactance. The transformer ratio can be given by:

$$\alpha = \frac{N_1}{N_2} = \frac{E_2}{E_1} \approx \frac{V_1}{V_2} \tag{3.12}$$

First, consider a practical transformer with its secondary terminals open-circuited. As in the case of the ideal transformer, a magnetizing current I_m is required to set up the mutual flux Φ and under linear magnetization conditions, it will be in phase with I_m. A practical transformer will obviously have iron loss in the core and corresponding core-loss component of current I_{e+h} in phase with E_1, will present these losses. The phasor sum of I_m and I_{e+h} is equal to the primary current at no-load, I_o.

Next, consider the practical transformer on load. The input primary current I_1 must now meet two requirements. It must supply the no-load current I_o of the and supply a current to stand for the required secondary current I_2. The voltage V_1 applied to the primary winding terminals can then be equated to the phasor sum of the voltage drops in the primary resistance R_1 and the primary leakage reactance X_1 and the primary EMF E_1. For the secondary winding the transformation ratio is taken into account by introducing an ideal transformer ratio E_1/E_2. The secondary EMF E_2 can then be equated to the phasor sum of the secondary voltage V_2 and the voltage drops in the secondary resistance R_2 and leakage reactance X_2. Figure 3.5 presents the phasor diagram of a transformer under no load conditions.

So, the practical transformer will be presented by an ideal transformer, which relates E_1 and E_2 in addition to two separate electrical circuits which represent the primary the primary and secondary windings as shown in Figure 3.6.

However, from Figure 3.6 and by applying the ordinary AC circuit theory, it is possible to represent the practical transformer by the following phasor equations:

$$V_1 = E_1 + I_1 Z_1 \tag{3.13}$$

$$E_2 = V_2 + I_2 Z_2 \tag{3.14}$$

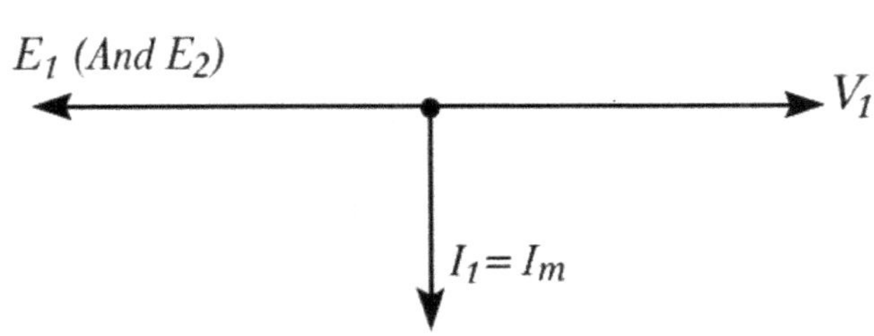

FIGURE 3.5 Phasor diagram of a transformer on no load.

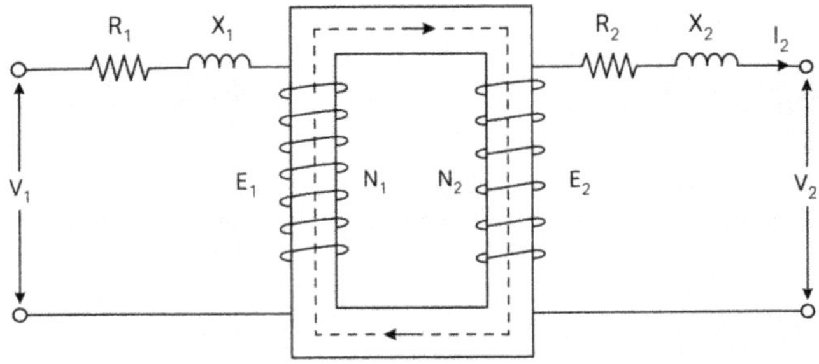

FIGURE 3.6 Representation of real transformer.

$$Z_1 = R_1 + j\,X_1 \tag{3.15}$$

$$Z_2 = R_2 + j\,X_2 \tag{3.16}$$

Example 3.5: A 2,300 to 240 V step-down transformer is rated at 150 kVA and 50 Hz. Its windings have the following resistance and inductance: $R_l = 0.1\ \Omega$, $X_l = 0.3\ \Omega$, $R_2 = 0.001\ \Omega$, $X_2 = 0.003\ \Omega$. When the transformer is operating at rated load of 0.8 p.f. lagging the supply is feeding the transformer at 0.75 p.f. lagging. Calculate:

(a) Primary and secondary current
(b) Primary and secondary winding impedances
(c) Primary and secondary winding voltage drops
(d) Primary and secondary induced voltages
(e) Total turns of the low voltage winding
(f) The transformation ratio
(g) Ratio of terminal voltages.

SOLUTION:

(a) $I_1 \dfrac{V_A}{V_1} = \dfrac{150\times10^3}{2300} = 65.2\ \angle -41^\circ\ \text{A}$

$I_2 \dfrac{V_A}{V_2} = \dfrac{150\times10^3}{240} = 625\angle -37^\circ\ \text{A}$

(b) $Z_1 = R_1 + j\,X_1\ 0.1\ +\ \text{j}\ 0.3\ = 0.316\ \angle 71.5^\circ\ \Omega$

$Z_2 = R_2 + j\,X_2\ = 0.001\ +\ \text{j}\ 0.003 = 0.00316\ \angle 71.5^\circ\ \Omega$

(c) $I_1\ Z_1 = 65.2\ \angle -41^\circ \times 0.316\ \angle 71.5^\circ = 20.6\ \angle 30.5^\circ\,\text{V}$

$I_2\ Z_2 = 625\ \angle -37^\circ \times 0.00316\ \angle 71.5^\circ = 1.97\ \angle 34.5^\circ\,\text{V}$

(d) $E_1 = V_1 - I_1 Z_1 = 2300 - 20.6 \angle 30.5^\circ = 2282.3 \angle -0.26^\circ \text{V}$

$E_2 = V_2 + I_2 Z_2 = 240 + 1.97\angle 34.5^\circ = 241.6\angle 0.26^\circ \angle 0.26^\circ \text{V}$

(e) The transformation ratio

$$\alpha = \frac{E_2}{E_1} = 9.44$$

(f) Ratio of terminal voltage

$$\frac{V_1}{V_2} = 10$$

3.6 EQUIVALENT CIRCUIT FOR A PRACTICAL TRANSFORMER

The term "equivalent circuit of a machine" means the combination of fixed and variable resistances and reactance, which exactly simulates performance and working of the machine.

For a transformer, no load primary current has two components,

$$I_m = I_o \sin \Phi_o = \text{Magnetizing component}$$

$$I_{e+h} = I_o \cos \Phi_o = \text{Active component}$$

I_m produces the flux and is assumed to flow through reactance X_m called no load reactance while I_{e+h} is active component representing core losses hence is assumed to flow through the reactance R_{e+h}. Hence equivalent circuit on no load can be shown as in the Figure 3.7. This circuit consisting of R_{e+h} and X_m in parallel is called exciting circuit. From the equivalent circuit we can write, $R_{e+h} = V_1/I_{e+h}$ and $X_m = V_1/I_m$.

When the is connected to the transformer then secondary current I_2 flows. This causes voltage drop across X_2 and R_2. Due to I_2, primary draws an additional current.

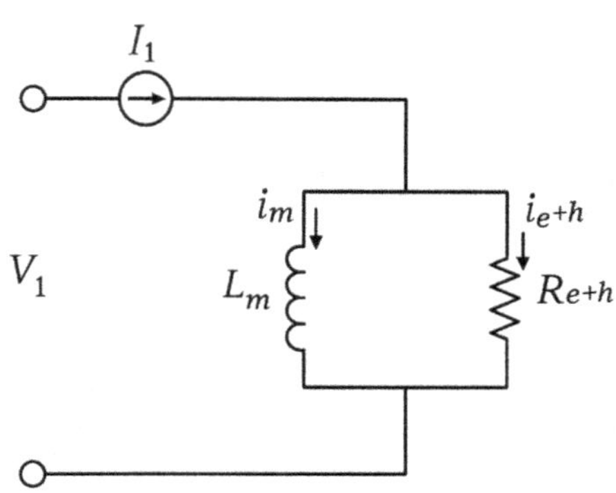

FIGURE 3.7 No-load equivalent circuit.

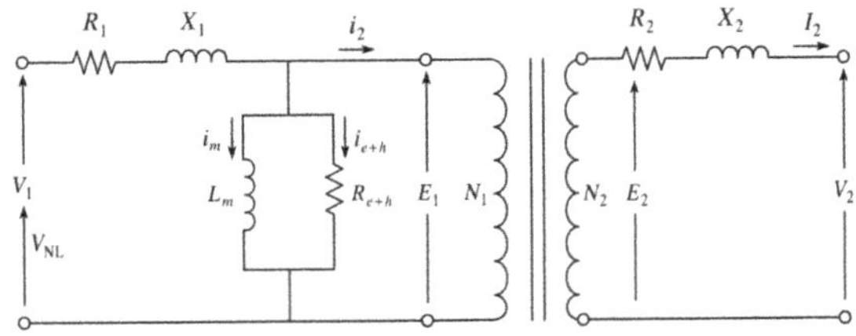

FIGURE 3.8 Equivalent circuit of real transformer.

$I_2' = I_{2}/\alpha$. Now I_l is the phasor addition of I_o and I_2'. This I_l causes the voltage drop across primary resistance R_1 and reactance X_l. Hence, the equivalent circuit can be shown as in the Figure 3.8.

3.6.1 Derivation of the Equivalent Circuit

Consider the secondary voltage equation (3.14), to be:

$$E_2 = V_2 + I_2\left(R_2 + j\,X_2\right) \tag{3.17}$$

Since

$$E_2 = \frac{E_1}{\alpha} \quad \text{and} \quad I_2 = \alpha I_2'$$

(ideal transformer conditions) equation (3.17) can be written in the form

$$\frac{E_1}{\alpha} = V_2 + \alpha I_2'\left(R_2 + j\,X_2\right) \tag{3.18}$$

$$E_1 = \alpha V_2 + I_2'\left(\alpha^2 R_2 + j\,\alpha^2 X_2\right) \tag{3.19}$$

Thus, the equivalent circuit of Figure 3.8 can be redrawn as shown in Figure 3.9, in which $R_2' = \alpha^2 R_2$ and $X_2' = \alpha^2 X_2$ are, respectively, the secondary resistance and leakage reactance referred to the primary and $V_2' = \alpha\, V_2$ is the secondary terminal voltage referred to the primary. It will be more convenient no to change I_2 into I_2' which stands for the secondary load current referred to primary.

The circuit of Figure 3.9 is usually drawn with the ideal transformer omitted as shown in Figure 3.10, and this circuit is usually known as the exact equivalent circuit.

3.6.2 The Approximate Transformer Equivalent Circuit

Transformers which are used at a constant power frequency can have very simplified approximate equivalent circuits, without having a substantial effect on the performance evaluation (efficiency and voltage regulation). It should be borne in mind that

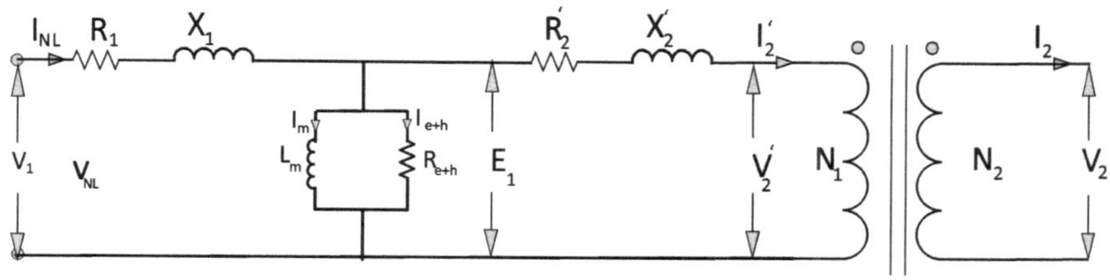

FIGURE 3.9 Equivalent circuit with referred values.

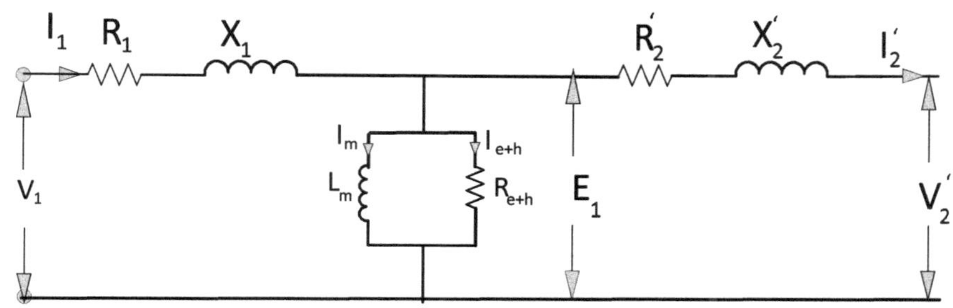

FIGURE 3.10 Exact equivalent circuit of a practical transformer.

higher the VA or kVA rating of the transformers, better are the approximation-based evaluation results.

It is assumed that $E_1 \approx V_1$ (V_1 is approximately equal to E_1) even under conditions of load. This assumption is justified because the values of winding resistance and leakage reactances are very small. Therefore, the exciting current drawn by the parallel combination of conductance G_{e+h} and susceptance B_m would not be affected significantly by shifting it to the input terminals. With this change, the equivalent circuit becomes as shown in Figure 3.11.

Denoting
$$R_{eq1} = R_1 + R_2' \tag{3.20}$$

and
$$X_{eq1} = X_1 + X_2' \tag{3.21}$$

The equivalent circuit becomes as shown in Figure 3.12. R_{eq1} and X_{eq1} are called the equivalent resistance and equivalent reactance referred to primary side.

In the previous approximate equivalent circuit, the ideal transformer is used to transfer voltages, currents and impedances from the secondary side to the primary side. In this case, the circuit represents the transformer looking into the load terminals, may be of great importance. In this case, the ideal transformer is used to transfer voltages, currents and impedances from the primary side to the secondary side which results in the circuits shown in Figure 3.13.

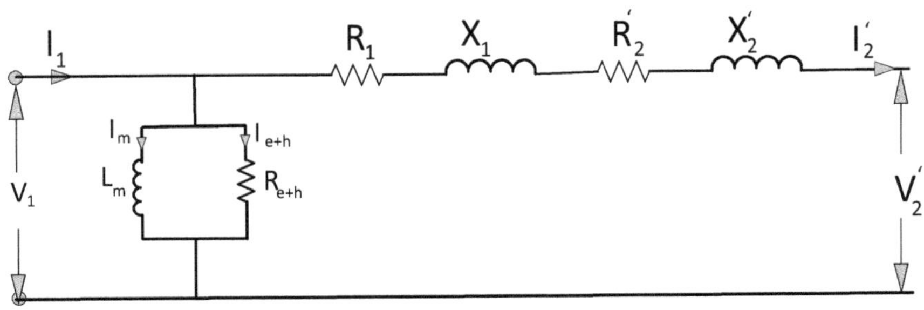

FIGURE 3.11 Equivalent circuit referred to primary side.

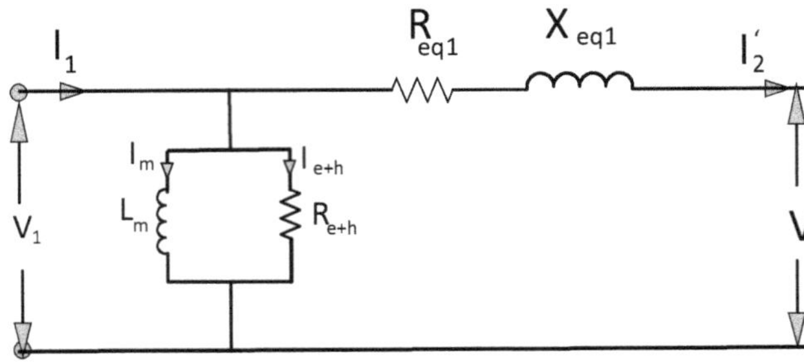

FIGURE 3.12 Approximate equivalent circuit.

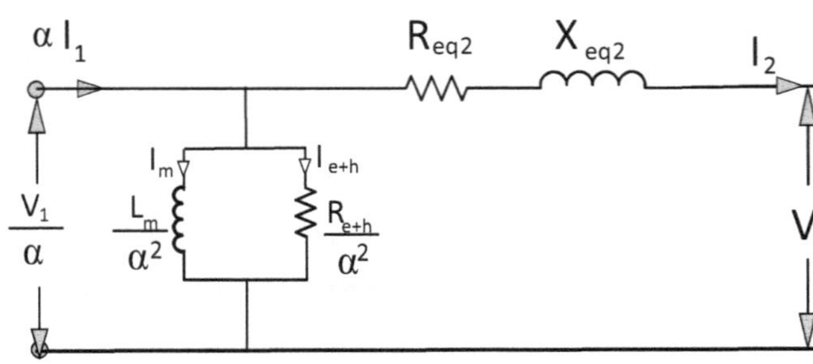

FIGURE 3.13 Approximate transformer equivalent circuit referred to the secondary.

The new series impedance shown in Figure 3.13 is given by

$$Z_{eq2} = R_{eq2} + j\,X_{eq2} \tag{3.22}$$

where,

$$R_{eq2} = R_2 + \frac{R_1}{\alpha^2} \tag{3.23}$$

$$X_{eq2} = X_2 + \frac{X_1}{\alpha^2} \tag{3.24}$$

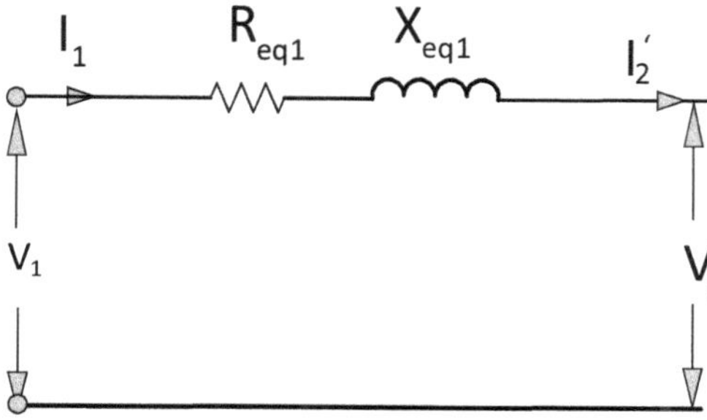

FIGURE 3.14 Most simplified form of approximate equivalent circuit.

However, in most power systems computations, the transformer is presented by a series reactance only since the effect of their R_{eq} or the shunt branch I_o is very small quantity. If only voltage regulation is to be calculated, even the excitation branch can be neglected and the equivalent circuit becomes as shown in Figure 3.14.

Example 3.6: A transformer has 600 primary turns and 150 secondary turns. The primary and secondary resistances are 0.25 Ω and 0.01 Ω, respectively, and the corresponding leakage reactances are 1.0 Ω and 0.04 Ω, respectively. Determine (a) the equivalent resistance referred to the primary winding, (b) the equivalent reactance referred to the primary winding, (c) the equivalent impedance referred to the primary winding, and (d) the phase angle of the impedance.

Solution:

(a) From equation (3.20), equivalent resistance

$$\mathrm{R}_{eq1} = R_1 + \alpha^2 R_2$$

$$\mathrm{R}_{eq1} = 0.25 + \left(\frac{600}{150}\right)^2 0.01 = 0.41\,\Omega$$

(b) $X_{eq1} = X_1 + \alpha^2 X_2$

$$\mathrm{R}_{eq1} = 1.0 + \left(\frac{600}{150}\right)^2 0.04 = 1.64\,\Omega$$

(c) $Z_{eq1} = \sqrt{R_{eq1}^2 + X_{eq1}^2} = \sqrt{0.41^2 + 1.64^2} = 1.69\,\Omega$

(d) $\cos\phi = \dfrac{\mathrm{R}_{eq1}}{Z_{eq1}} = \dfrac{0.41}{1.69}$

Hence $\phi = \cos^{-1}\dfrac{0.41}{1.69} = 75.96^0$

Example 3.6: A 15 kVA, 2,200/110 V transformer has $R_1 = 1.75\ \Omega$, $R_2 = 0.0045\ \Omega$ the leakage reactances are $X_1 = 2.6\ \Omega$ and $X_2 = 0.0075\ \Omega$ Calculate,

(a) Equivalent resistance referred to primary.
(b) Equivalent resistance referred to secondary.
(c) Equivalent reactance referred to primary.
(d) Equivalent reactance referred to secondary.
(e) Equivalent impedance referred to primary.
(f) Equivalent impedance referred to secondary.
(g) Total copper loss.

Solution:

The given values are, $R_1 = 1.75\ \Omega$, $R_2 = 0.0045\ \Omega$, $X_1 = 2.6\ \Omega$, $X_2 = 0.0075\ \Omega$

$$\alpha = 110/2200 = 1/20 = 0.05$$

(a) $R_{eq1} = R_1 + R_2' = R_1 + R_2/\alpha^2 = 1.75 + 0.0045/0.05^2 = 3.55\Omega$

(b) $R_{eq2} = R_2 + R_1' = R_2 + \alpha^2 R_1 =$

$$= 0.0045 + (0.05)^2 \times 1.75 = 0.00887\ \Omega$$

(c) $X_{eq1} = X_1 + X_2' = X_1 + X_2/\alpha^2 = 2.6 + 0.0075/(0.05)^2 = 5.6\ \Omega$

(d) $X_{eq2} = X_2 + X_1' = X_2 + \alpha^2 X_1$

$$= 0.0075 + (0.05)^2 \times 2.6 = 0.014\ \Omega$$

(e) $Z_{eq1} = R_{eq1} + j\,X_{eq1} = 3.55 + \mathrm{j}\ 5.6\ \Omega$

$$Z_{eq1} = \sqrt{}(3.55^2 + 5.6^2) = 6.6304\ \Omega$$

(f) $Z_{eq2} = R_{eq2} + j\,X_{eq2} = 0.00887 + \mathrm{j}\ 0.014\ \Omega$

$$Z_{eq2} = \sqrt{}(0.00887^2 + 0.014^2) = 0.01657\ \Omega$$

(g) To find the load copper loss, calculate full load current.

$$I_{1F.L.} = (\text{kVA} \times 1000)/V_1 = (25 \times 1000)/2200 = 11.3636\ \text{A}$$
$$\text{total copper loss} = (I_{1F.L.})^2\ R_{eq1} = (11.3636)^2 \times 355 = 458.4194\ \text{W}$$

This can be checked as,

$$I_{1F.L.} = (\text{kVA} \times 1000)/V_2 = (25 \times 1000/110 = 227.272\ \text{A}$$

Total copper loss $= I_1^2R_1 + I_2^2R_2$

$$= (11.3636)^2 \times 1.75 + (227.373)^2 \times 0.0045$$

$$= 225.98 + 232.4365 = 458.419 \text{ W}$$

3.7 DETERMINATION OF EQUIVALENT CIRCUIT PARAMETERS

The equivalent circuit shown in Figure 3.11 can be used to predict the performance of the transformer. All the circuit parameters must be known so that the equivalent circuit can be used for the above purpose. These parameters can be easily determined by performing tests that involve little power consumption. Two tests, a no-load test (or open circuit test) and short circuit test, will provide information for determining the parameters of the equivalent circuit.

3.7.1 Open Circuit (OC) Test

The shunt branch parameters can be determined by performing this test. Since, the core loss and the magnetizing current depend on applied voltage, this test is performed by applying the rated voltage to one of the windings keeping the other winding open (generally HV winding is kept open and rated voltage is applied to LV winding). The circuit diagram to conduct this test is shown in Figure 3.15. Since, the secondary terminals are open (no load is connected across the secondary), current drawn from the source is called as no load current. On no-load, the approximate equivalent circuit shown in Figure 3.11 can be further reduced and is shown in Figure 3.13a.

Under no-load condition the power input to the transformer is equal to the sum of losses in the primary winding resistance R_1 and core loss. Since, no load current is very small, the loss in winding resistance is neglected. Hence, on no load, the power drawn from the source is dissipated as heat in the core. If I_o and P_i are the current and input power drawn by the transformer at rated voltage V_l respectively, then

$$\cos\phi_0 = \frac{P_i}{V_l I_o}$$

From Figure 3.13b,

$$I_{e+h} = I_0 \cos\varphi_0, \qquad I_m = I_0 \sin\varphi_0,$$

Therefore,

$$R_{e+h} = \frac{V1}{I_{e+h}}, \qquad X_m = \frac{V_1}{I_m},$$

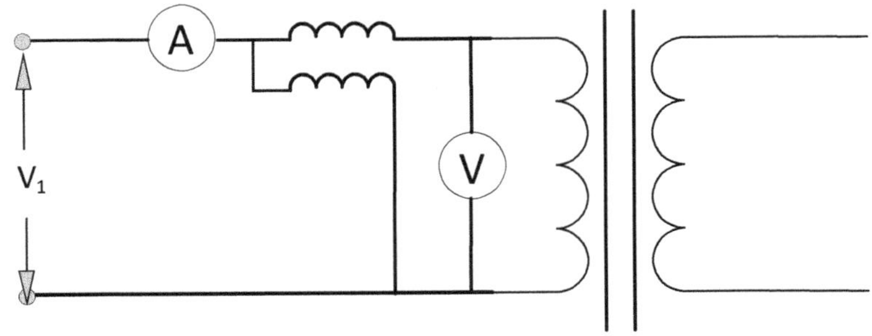

FIGURE 3.15 Circuit diagram at no-load test.

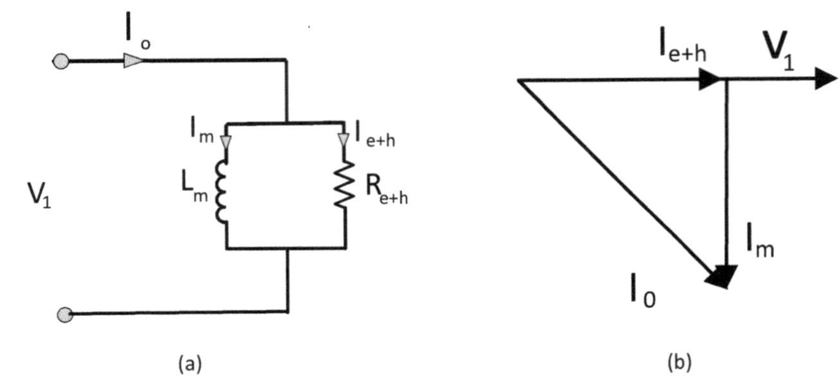

FIGURE 3.16 (a) Equivalent circuit at no load, (b) phasor diagram at no-load

3.7.2 Short Circuit (SC) Test

Consider the circuit shown in Figure 3.11. Suppose the input voltage is reduced to a small fraction of rated value and secondary terminals are short-circuited. A current will circulate in the secondary winding. Since a small fraction of rated voltage is applied to the primary winding, the flux in the core and hence the core loss is very small. Hence, the power input on SC is dissipated as heat in the winding. The circuit diagram to conduct this test is shown in Figure 3.17a. In this test, the LV terminals of the transformer are short-circuited. The primary voltage is gradually applied till the rated current flows in the winding. Since, the applied voltage is very small (may be of the order of 5–8 %), the magnetizing branch can now be eliminated from the equivalent circuit. The modified equivalent circuit is shown in Figure 3.17b. If V_{SC} is the applied voltage to circulate the rated current I_2' on SC, and Pc is the power input to the transformer then,

$$Z_{SC} = \frac{V_{SC}}{I_2'}, \qquad \cos\varphi_{sc} = \frac{P_c}{V_{sc} I_2'}$$

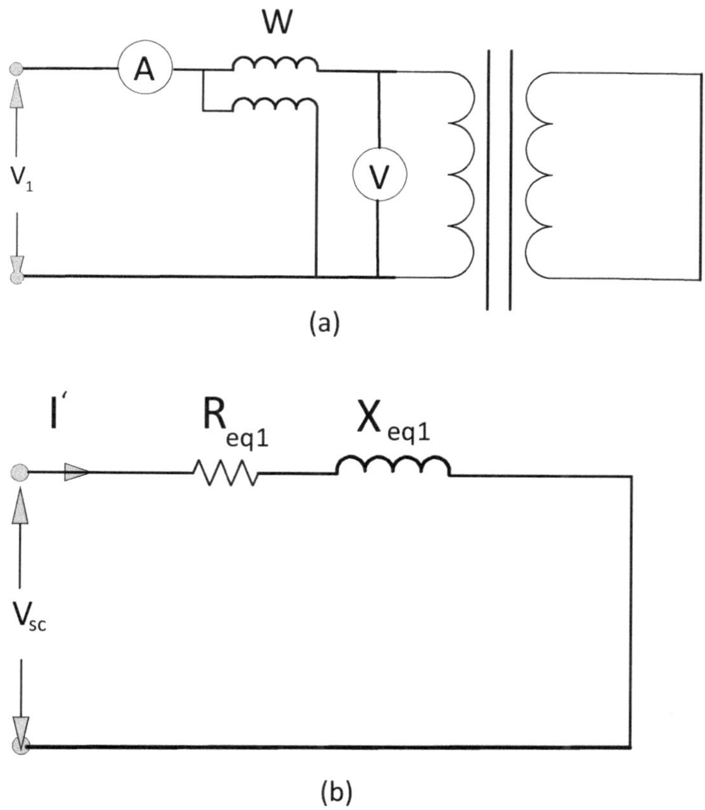

FIGURE 3.17 (a) Circuit diagram at short-circuit test, (b) Equivalent circuit at short circuit

Therefore,

$$R_1 + R_2' = Z_{sc}\cos\varphi_{sc}, \qquad X_1 + X_2' = Z_{sc}\sin\varphi_{sc},$$

Example 3.7: A 20 kVA, 2,500/250 V, 50 Hz, single-phase transformer gave the following test result:

Open-circuit test: 250 V, 1.5 A, and 100 W measured on the LV side
Short-circuit test: 100 V, 8 A, and 320 W measured on the on HV side

(a) Compute the parameters of the equivalent circuits and draw the equivalent circuit of the transformer showing all the values.

Solution: From the given test results data, we have

$$\mathrm{R}_{eq1} = \frac{P_{sc}}{I_{sc}^2} = \frac{320}{8^2} = 5\ \Omega$$

$$X_{eq1} = \sqrt{\left(\frac{V_{sc}}{I_{sc}}\right)^2 - 5^2} = 12\Omega$$

$$\mathrm{R}_{e+h} = \frac{V_{oc}^2}{P_{oc}} = \frac{250^2}{100} = 12\Omega$$

$$X_m = \sqrt{\left(\frac{V_{oc}}{I_{oc}}\right)^2 - 12^2} = 173\Omega$$

R_{e+h} and X_m are LV side values, which can be converted to the HV side

$$\mathrm{R}_{e+h} = R_{e+h} * \left(\frac{N_1}{N_1}\right)^2 = 62.5 \text{ k } \Omega$$

$$X_m = X_m * \left(\frac{N_1}{N_1}\right)^2 = 17.3 \text{ k } \Omega$$

3.8 VOLTAGE REGULATION

Under no load conditions, the voltage at the secondary terminals is E_2, and $E_2 \approx V_1 \times \left(\frac{N_1}{N_2}\right)$.

(This approximation neglects the drop R_l and X_l due to small no load current). As load is applied to the transformer, the load current or the secondary current increases. Correspondingly, the primary current I_l also increases. Due to these currents, there is a voltage drop in the primary and secondary leakage reactances, and as a consequence, the voltage across the output terminals or the load terminals changes. In quantitative terms, this change in terminal voltage is called voltage regulation.

Voltage regulation of a transformer is defined as the drop in the magnitude of load voltage (or secondary terminal voltage) when load current changes from zero to full load value. The secondary rated voltage of a transformer is equal to the secondary terminal voltage at no load (i.e., E_2). Voltage regulation is generally expressed as a percentage:

$$VR\%(\text{percent voltage regulation}) = \frac{E_2 - V_2}{V_2} \times 100 \tag{3.25}$$

Note that the magnitude of the voltages, not their phasor values, are used and E_2 is the secondary no-load ($I_2 = 0$) while V_2 is the secondary voltage under load keeping the primary voltage constant. Referring to Figure 3.15, it will show that when there is no load connected to the secondary terminals of transformer, $I_2 = 0$ and secondary voltage will be

$$E_2 = \frac{V_1}{\alpha} \tag{3.26}$$

The regulation expression becomes

$$VR\%\,(\text{percent voltage regulation}) = \frac{\frac{V_1}{\alpha} - V_2}{V_2} \times 100 \tag{3.27}$$

Example 3.7: A 10 kVA, 2,400/240 V, single-phase transformer have the following parameters: $R_l = 3\Omega$, $R_2 = 15\Omega$ the leakage reactance are $X_l = 0.03\,\Omega$ and $X_2 = 0.15\,\Omega$. Calculate the full-load voltage regulation of the transformer when operated at rated secondary voltage and (a) 0.8 power factor lagging and (b) 0.8 power factor leading.

Solution:

$$\mathrm{R}_{eq2} = R_2 + R_l' = R_2 + R_l/\alpha^2 =$$

$$= 0.03 + 3/(10)^2 = 0.06\,\Omega$$

$$X_{eq2} = X_2 + X_l' = X_2 + X_l/\alpha^2 =$$

$$= 0.15 + 15/(10)^2 = 0.3\,\Omega$$

$$Z_{eq2} = R_{eq2} + j\,X_{eq2} = 0.06 + \mathrm{j}\ 0.3\ \Omega = 0.3\ \angle 78.7^0$$

(a) For a 0.8 power factor, lagging

$$I_2 = \frac{S}{V_2} = 41.7\ \mathrm{A}$$

Referring to Figure 10.17

$$\frac{V_1}{\alpha} = V_2 + I_2\,Z_{eq2} = 240\angle 0 + 41.7\angle -37^0 \times 0.3\ \angle 78.7^0$$

$$= 249.5\ +\ \mathrm{j}\ 8.5\ =\ 249.65\ \angle 1.95^0$$

$$VR\% = \frac{\frac{V_1}{\alpha} - V2}{V_2} \times 100 = \frac{249.65 - 240}{240} \times 100 = 4\%$$

(b) For a 0.8 power factor, leading

$$I_2 = 41.7\angle +37^0 \text{ A}$$

$$\frac{V_1}{\alpha} = V_2 + I_2 Z_{eq2} = 240\angle 0 + 41.7\angle +37^0 \times 0.3\angle 78.7^0$$

$$= 234.5 + \text{j } 11.5 = 235\angle 2.8^0$$

$$VR\% = \frac{249.65 - 240}{240} \times 100 = -2.18\%$$

3.9 LOSSES AND EFFICIENCY

In a transformer, there are two primary types of power losses:

1. Iron loss
2. Copper loss

1) **Iron Loss:** This occurs in the iron components of the transformer and is influenced by the alternating frequency of the electromagnetic force. Iron loss is subdivided into:
   i. **Eddy current loss:** This loss arises from the alternating flux that links with the core, inducing an EMF known as eddy EMF This induces eddy currents within the core that, due to the core's inherent resistance, convert into heat, termed eddy current loss. The magnitude of eddy current loss is proportional to the square of the supply frequency.

   ii. **Hysteresis loss:** This loss results from the repetitive magnetic reversal of the flux within the core, which generates heat. Hysteresis loss is directly proportional to the supply frequency.

   Methods to reduce iron losses include using a core made from thin laminations of silicon steel, each coated with varnish to impede eddy currents, and selecting materials with high permeability to decrease hysteresis loss.

2) **Copper loss:** Occurring in the transformer's primary and secondary coils when under load, this loss manifests as heat generated by the resistance of the coils. Copper loss varies with the load and is thus known as a variable loss, whereas iron loss remains constant as it depends on the fixed supply voltage and frequency.

Efficiency: It is the ratio of the output power to the input power of a transformer

Input = Output + Total losses = Output + Iron loss + Copper loss

$$\text{Efficiency } (\eta) = \frac{\text{output power}}{\text{output power + Iron loss + Copper loss}}$$

$$= \frac{v_2 I_2 \cos\phi}{v_2 I_2 \cos\phi + P_{iron} + P_{copper}}$$

where, v_2 is the secondary (output) voltage, I_2 is the secondary (output) current and $\cos \phi$ is the power factor of the load.

The transformers are normally specified with their ratings as kVA,

Therefore,

$$\text{Efficiency} = \frac{KVA\,(10^3)\,cos\,\phi}{KVA\,(10^3)\,cos\,\phi + P_{iron} + P_{Copper}}$$

Since the copper loss varies as the square of the load, the efficiency of the transformer at any desired load X is given by

$$\text{Efficiency} = \frac{X\,KVA(10^3)\,cos\phi}{X\,KVA(10^3)\,cos\phi + P_{iron} + P_{Copper}}$$

where P_{Copper} is the copper loss at full load $= I^2 R$ watts

Condition for Maximum Efficiency:

For any device to achieve maximum efficiency, its losses must be minimized. In the case of transformers, the iron loss represents a fixed loss, while the copper loss varies and is considered a variable loss. Maximum efficiency is attained when these two losses are not only equal but also minimized. Thus, the condition for achieving peak efficiency in a transformer can be expressed as:

$$\text{Iron loss} = \text{Copper loss}$$

where each is at its minimum level.

3.10 POLY-PHASE TRANSFORMERS

When three single-phase transformers with their primary windings connected to a 3-phase supply, as shown in Figure 3.18, the fluxes ϕ_1, ϕ_2 and ϕ_3 produced in the individual units will be equal in magnitude but will be displaced by 120° from each other.

Since at any instant the total flux $\phi_1 + \phi_2 + \phi_3 = 0$, then when the three transformers are placed side by side with a common yoke top and bottom, there is no need to provide a return path for the flux. We have thus a 3-limb transformer of the core types.

The common types of the 3-phase transformers are the core type and the shell type. The lengths of the magnetic circuits of the 3-phase are not quite equal, since the outer limbs include a portion of the yoke. Therefore, the magnetizing currents of the 3-phases are never exactly in balance.

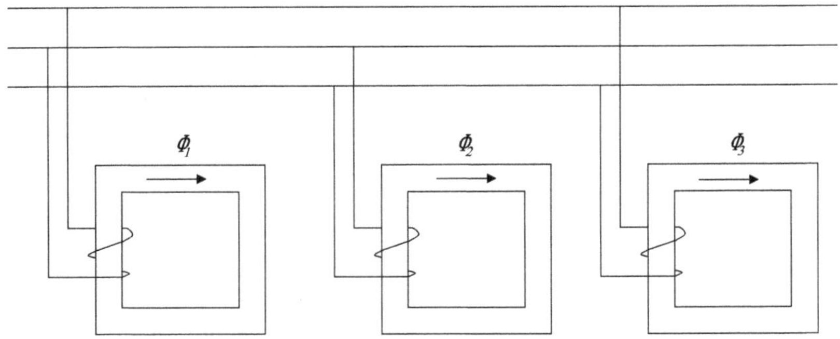

FIGURE 3.18 Connection of three single–phase transformers.

3.10.1 Winding Connections

The stander lettering for the phases is A, B, and C. Capital letters are used for the HV winding and lower – case letters a, b, and c for LV windings. Each winding has two terminals designated by subscript number 1 and 2. The polarity of a transformer is designated as subtractive when the leads A_1 and a_1 are adjacent, or as additive when A_1 and a_1 are located diagonally opposite.

To test for polarity, connect an adjacent HV and LV terminal to gather as shown in Figure 3.20. Apply an AC Voltage to the HV terminals A_1 A_2. Measure the voltage between the other adjacent HV and LV leads. If this greater than the voltage applied to A_1 A_2 the polarity is additive if less is subtractive.

3.10.2 Three-Phase Connections for Transformers

Since most power systems are 3-phase, it is necessary to provide 3-phase transformer connections. The main connections are:

(a) star-star	(Y-Y)	called	Y_y
(b) delta-delta	(Δ-Δ)	called	D_d
(c) star-delta	(Y-Δ)	called	Y_d
(d) delta-star	(Δ-Y)	called	D_y
(e) star-zigzag	(Y-Z)	called	Y_z
(f) delta-zigzag	(Δ-Z)	called	D_z

In discussion the various connection, the ratio of all transformers will be assumed equal, and the magnetizing current and impedances of the transformers will be neglected in the vector diagram for a balanced load.

A vector diagram can be constructed on the following general principles:

(i) The voltages of corresponding primary and secondary windings on the same limp (i.e., the input or applied primary voltage, and the developed secondary output V) are in phase opposition, and the two induced EMFs are in phase.

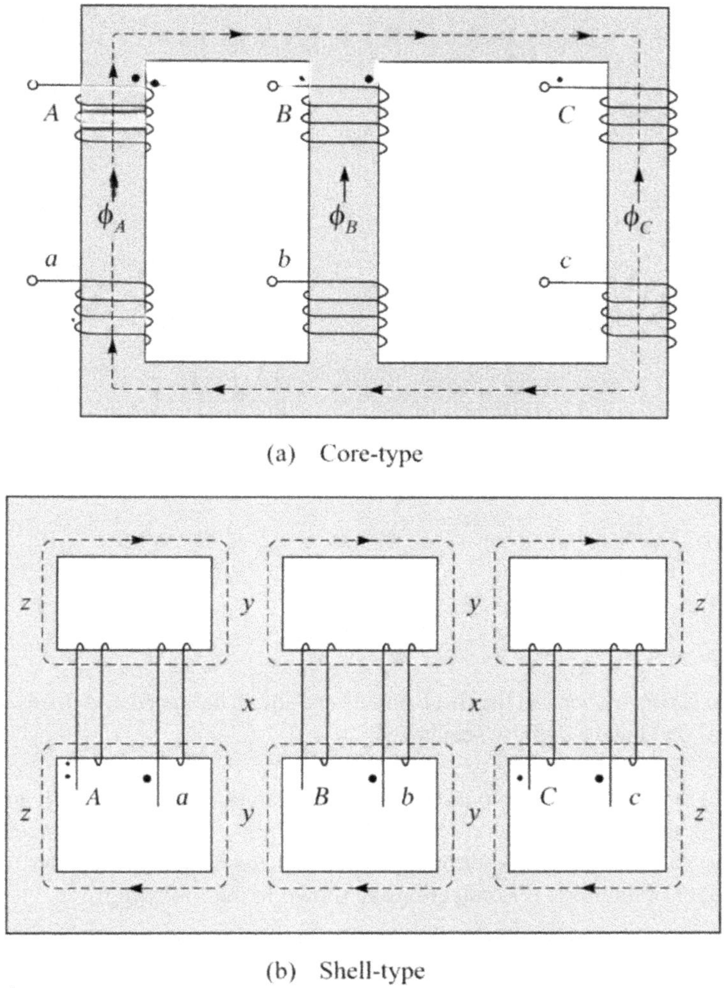

FIGURE 3.19 Three-phase transformers.

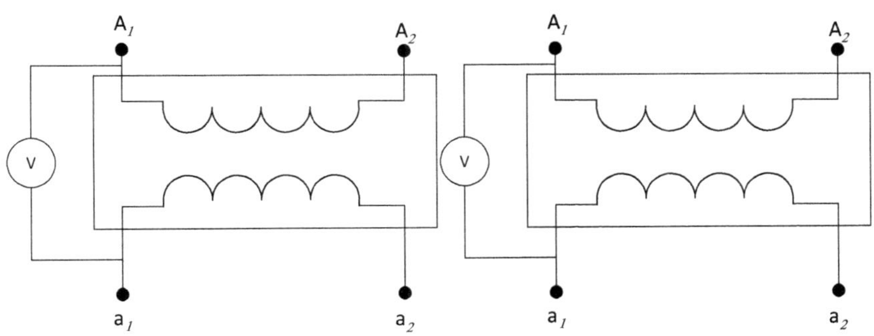

FIGURE 3.20 Test for polarity.

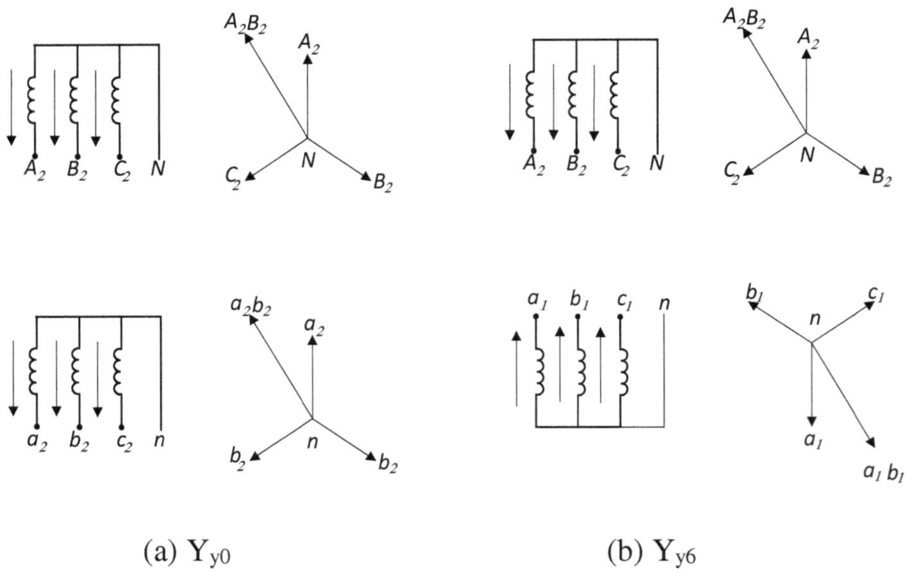

FIGURE 3.21 Star-star connection.

(ii) The EMFs induced in the three phases are equal, balanced and displaced by 120° and have a definite sequence.

If the induced EMF in an HV phase A_1A_2 be in direction A_1 to A_2 at a given instant, then the induced EMF in the corresponding LV. phase at the same instant will be from a_1 to a_2. The vector diagrams represent induced EMFs (not applied voltages), for a number of methods of connections as shown in the following.

3.10.2.1 The Star-Star Connection Y_y

For the connection Y_y, the phase difference between the secondary voltage of any phase and the corresponding phase EMF is zero (see Figure 3.21a). But for the second connection Y_{y6}, the phase difference between the secondary EMF of any corresponding phase EMF is 180°, as shown in Figure 3.21b.

The above connections are the most economical connections for small, HV transformers, as the number of turns per phase and the amount of insulation is a minimum $(V_{ph} = V_l/\sqrt{3})$. Also, the possibility of utilizing both star points for a fourth wire may be useful.

If the transformer is supplied with a sinusoidal voltage, it will take an exciting current having third harmonic component. With the neutral point N connected to that of the supply, these third harmonic component of current being in phase, and have a return path through the neutral connection. In absence of the neutral point the third harmonic of the exciting current distort the flux wave and cause a distortion in the secondary voltage.

3.10.2.2 The Delta-Delta Connection D_d

This is an economical connection for large power, LV transformers ($I_{ph} = I_l / \sqrt{3}$). Although it increases the number of turns per phase, but reduces the necessary sectional area of conductors.

The closed mesh serves to damp out the third harmonic currents. Therefore, the secondary voltage is sinusoidal. The absence of a star-point may be disadvantage.

3.10.2.3 The Star-Delta Connection Y_d

The star-delta arrangement is very common for power supply transformer. It has the advantage of star-point in HV side, and a delta winding to carry third harmonic currents, which stabilize the star-point potential. Also, when the HV winding is star-connected, there is some saving in cost of insulation.

3.10.2.4 The Delta-Star Connection D_y

This connection is very propeller for distributed transformers. It has the advantage of star secondary when it is desired to work motors and lighting from a four-wire LV side.

3.10.2.5 The Star-Zigzag Connection Y_z

The interconnection between phases effects a reduction of third harmonic voltages and at the same time permits of unbalanced loading since the phase voltages are composed on the zigzag-side of two half-voltages with a phase difference of 60°, 15% more turns are required for a given total voltage per phase compared with a normal phase connection. This connection is suitable for distributed transformers.

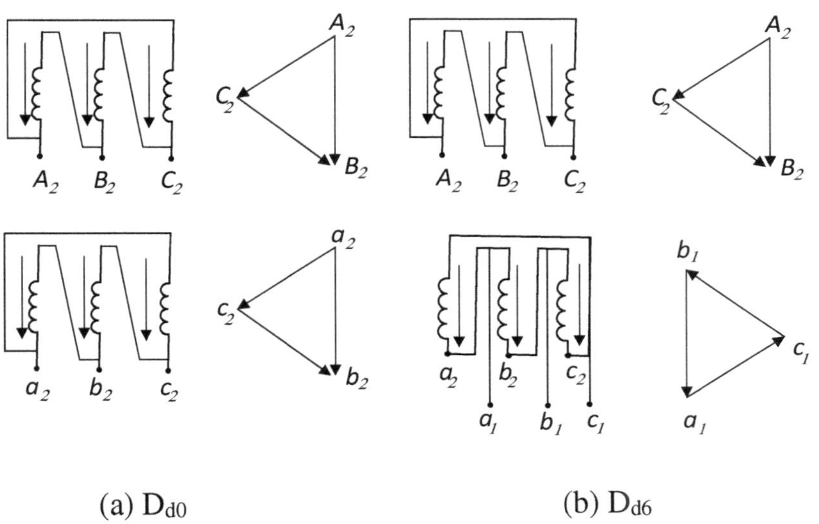

FIGURE 3.22 Delta-delta connection.

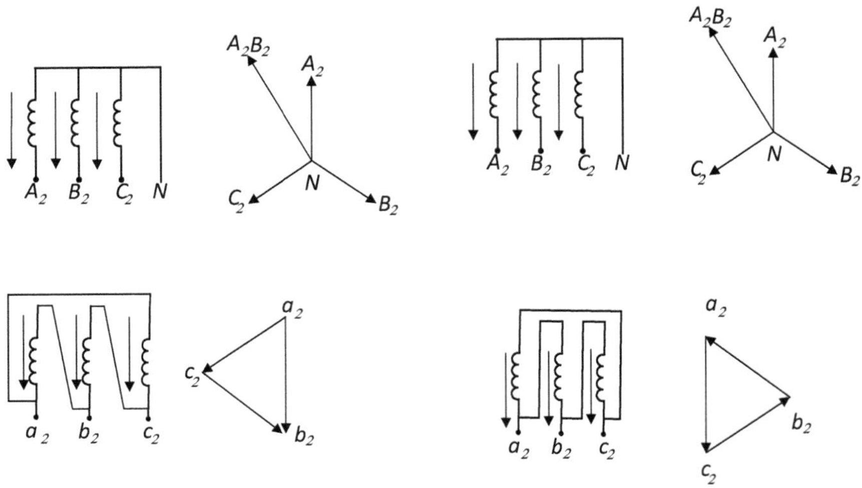

(a) Yd1 (b) Yd11

FIGURE 3.23 Star-delta connection.

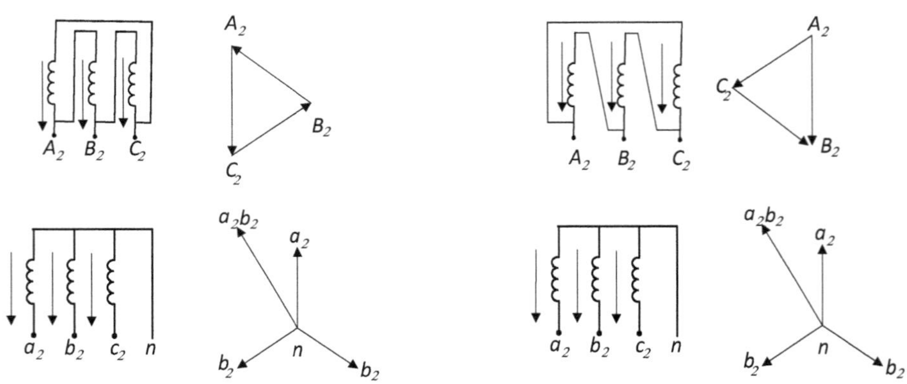

(a) Dy1 (b) Dy11

FIGURE 3.24 Delta-star connection.

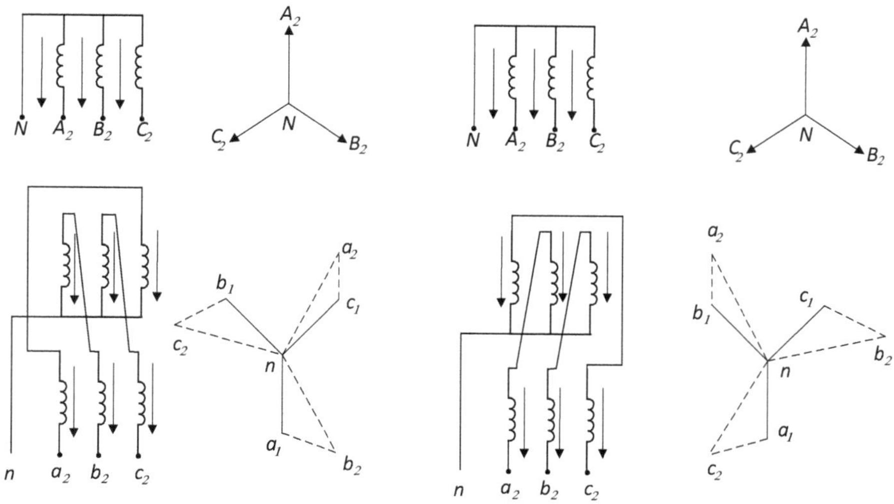

(a) Y_{z1} (b) Y_{z11}

FIGURE 3.25 Star-zigzag connection.

3.10.2.6 The Delta-Zigzag Connection D_z

This connections have the advantage of primary delta connection (in primary) and the advantage of secondary zigzag connection.

From previous, the groups into which all – possible 3-phase transformer connections are classified, are given as

Group 1	zero phase displacement	(Y_{y0}, D_{d0}, D_{z0})
Group 2	180° phase displacement	(Y_{y6}, D_{d6}, D_{z6})
Group 3	30° lag phase displacement	(Y_{d1}, D_{y1}, Y_{z1})
Group 4	30° lead phase displacement	$(Y_{d11}, D_{y11}, Y_{z11})$

This classifications are based on the phase angle between the primary and secondary induced EMFs (or the phase angle of the secondary when the primary is taken as a reference). The transformers of the same group can be operated in parallel (zero phase shifted).

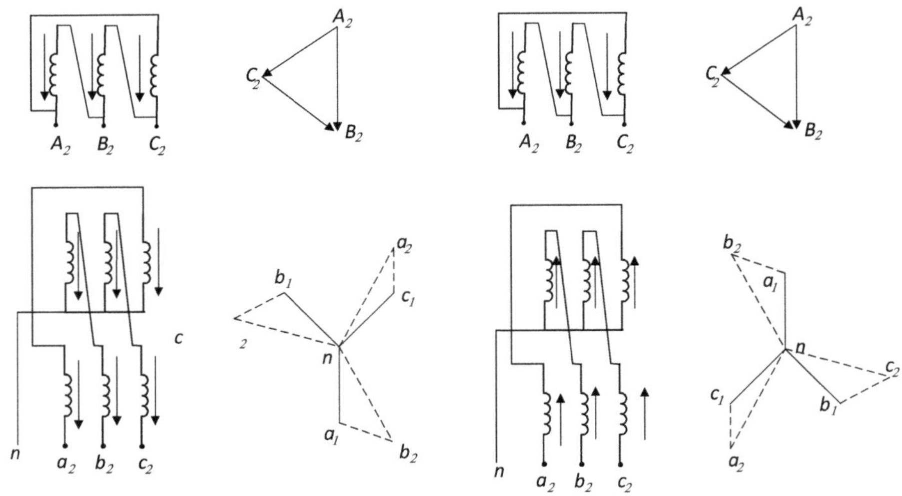

(a) D_{z0} (b) D_{z6}

FIGURE 3.26 DELTA-ZIGZAG CONNECTION.

Example 3.8: The constants of a 150 kVA, 2,400/240 V, 60 Hz transformer are as follows:

Resistance of the 2,400 V winding $R_1 = 0.216$ ohm

Resistance of the 240 V winding $R_2 = 0.00210$ ohm Leakage reactance of the 2,400 V winding $X_1 = 0.463$ ohm

Leakage reactance of the 240 V winding $X_2 = 0.00454$ ohm

Exciting admittance on the 240 V side $y_0 = 0.0101 - j0.069$ mho

Determine: (a) The values of the leakage impedances and the exciting admittance expressed in complex form.

(b) The efficiency of the transformer at rated load 0.80 power factor.

(c) The efficiency of the transformer at one-half rated load 0.60 power factor.

SOLUTION:

The rated secondary terminal voltage is 240 V and the rated load (rated output) of the transformer is 150,000 VA at 240 V. The rated load current is, therefore,

$$I_L = \frac{150{,}000}{240} = 625 \text{ A}$$

Let the phasor that represents the secondary terminal voltage V_2 lie on the axis of reals

$$V_2 = 240 + j\,0$$

and the load current expressed as a phasor is

$$I_L = 625(0.8\text{-j}0.6) = 500 - \text{j}\,375 \text{ A}$$

The secondary induced voltage E_2 must equal the phasor sum of the secondary terminal voltage plus the voltage drop across the secondary leakage impedance, thus

$$E_2 = V_2 + I_L\left(R_2 + \text{j}\,X_2\right)$$

$$= 240 + \text{j}\,0 + (500 - \text{j}\,375)\ (0.0021 + \text{j}\,0.00454)$$

$$= 240 + \text{j}\,0 + 1.05 + 1.70 + \text{j}\,2.27 - \text{j}\,0.79$$

$$= 242.75 + \text{j}\,1.48 = 242.8\ \angle 0.35^0$$

Since the primary induced voltage E_1 and the secondary induced voltage E_2 are both produced by the mutual flux, their ratio E_1/E_2 must equal the turns ratio $a = N_1/N_2 = 10$, and we have

$$E_1 = a\,E_2 = 10 \times 242.8\ \angle 0.35^\circ = 2428\ \angle 0.35^0 = 2428 + \text{j}\,14.8$$

The load component I_L' of the primary current expressed in phasor form is

$$I_L' = \frac{I_L}{a} = \frac{500 - j\,375}{10} = 50.0 - \text{j}\,37.5 \text{ A}$$

The primary current is the phasor sum of the load component I_L' and the exciting current I_0. The exciting current produces the mutual flux, so to speak, and is obtained by multiplying the primary induced voltage by the exciting admittance referred to the primary side, that is,

$$I_0 = E_1 \text{Y}_1$$

The value of the admittance in the given data is as measured on the secondary side. This admittance is referred to the primary side by making use of the impedance ratio a^2. It should be remembered that an impedance is transferred from the secondary side to the primary side of a transformer by multiplying its value by the impedance ratio. Further, impedance is the reciprocal of admittance, and on transferring the exciting admittance from the secondary to the primary side, we have

$$\frac{1}{Y_1} = a^2 \frac{1}{Y_2}$$

or

$$Y_1 = a^2 Y_2 = \frac{0.0101 - j0.069}{10^2}$$

$$= (1.01\ - \text{j}\,6.9)\ 10^{-4} = 6.98 \times 10^{-4} \angle -81.67^0$$

The primary exciting current is, therefore,

$$I_0 = E_l Y_1 = 6.98 \times 10^{-4} \angle -81.67^0 \times 2,428 \angle 0.35^0$$
$$= 1.70 = \angle -81.32^0 = 0.257 - \mathrm{j}\ 1.68$$

It is evident that the primary current is

$$I_1 = I_L^{'} + I_0$$
$$= 50.0 - \mathrm{j}\ 37.5 + 0.257 - \mathrm{j}\ 1.68$$
$$= 63.6 \angle -37.9^0 \angle -37.9^0$$

The primary applied voltage is the sum of the primary induced voltage and the voltage drop across the primary leakage impedance. Hence,

$$V_l = E_l + I_L (R_l + \mathrm{j}\ X_l)$$
$$= 2428 + \mathrm{j}\ 14.8 + (50.257 - \mathrm{j}\ 39.18)(0.216 + \mathrm{j}\ 0.463)$$
$$= 2456.99 + \mathrm{j}\ 29.5 = 2457 \angle 0.69^0$$

It is important to note that the transformer delivers rated load when it delivers its rating, at rated voltage, in VA, regardless of power factor.

$$\text{(b) Output} = 150000 \times 0.80 = 120000\ \mathrm{W}$$

Rated current = 62.5 A (on the HV side)

The equivalent resistance of the transformer referred to the HV side from is

$$R_{eq\,l} = 0.425\ \text{ohm}$$

and the load loss at rated current is

$$I^2{}_l R_{eq\,l} = (62.5)^2 \times 0.425 = 1660\ \mathrm{W}$$

which is also the value measured in the short-circuit test at rated current.

The no-load loss at rated voltage is

$$P_{oc} = 580\ \mathrm{W}$$

The total loss then is $1660 + 580 = 2240$ W

and the power input = output + losses

$$= 120000 + 2240 = 122240\ \mathrm{W}$$

The rated-load efficiency at 0.80 power factor is therefore

$$\eta = 1 - \frac{losses}{input} = 1 - \frac{2,240}{122,240}$$

$$\eta = 1 - \frac{losses}{input} = 1 - \frac{2,240}{122,240}$$

$$= 1\text{- }0.0183 = 0.9817$$

(c) One-half rated load $= 150000 \times 1/2 = 75000$ VA

$$\text{Output} = 75000 \times 0.60 = 45000 \text{ W}$$

The core loss is assumed to remain unchanged as the secondary voltage is again at rated value. Hence

$$P_{oc} = 580 \text{ W}$$

The load losses, however, vary as the square of the current and, therefore, at one-half rated current, are equal to one-fourth the load losses at full-load current. Hence

$$I^2 R_{eq} = 1/4 \times 1660 = 415 \text{ W}$$

and the total losses are $580 + 415 = 995$ W. The efficiency, then, at one-half rated load 0.60 power factor, is

$$\eta = 1 - \frac{losses}{input} = 1 - \frac{995}{45,000 + 995}$$

$$= 1 - \frac{995}{45,000} = 0.9784$$

Example 3.9: A source which can be represented by a voltage source of 8 V rms in series with an internal resistance of 2 kΩ is connected to a 50 Ω load resistance through an ideal transformer. Calculate the value of turns ratio for which maximum power is supplied to the load and the corresponding load power?

SOLUTION:

For maximum power transfer, the load resistance (referred to the primary) must be equal to the source resistance.

$$R_L' = \left(\frac{N_1}{N_2}\right)^2 R_L = n^2 R_L = 2000$$

$$n = \sqrt{\frac{2000}{50}} = 6.32$$

The primary current: $I_1 = \dfrac{V_s}{2R_s}$

Power supplied to the load: $P_{load} = R_L^{'} I_L^2 = 8$ mW

Example 3.10: A 460 V:2,400 V transformer has a series leakage reactance of 37.2 Ω as referred to the HV side. A load connected to the LV side is observed to be absorbing 25 kW, unity power factor, and the voltage is measured to be 450 V. Calculate the corresponding voltage and power factor as measured at the HV terminals.

Solution:

Secondary current:

$$I_2 = \frac{P_{load}}{V_{load}} = \frac{25000}{450} = 55.55 \text{ A}$$

Secondary current:

$$I_1 = \frac{460}{2400} \times 55.55 = 10.65 \text{ A}$$

Primary voltage:

$$V_1 = V_2^{'} + (j37.2)I_1$$

$$V_2^{'} = \frac{2400}{460} \times 450 = 2347.8 \text{ V}$$

$$V_1 = V_2^{'} + (j37.2)I_1 = 2381\angle 9.58^0$$

Power factor at primary terminals: $\cos(9.58^\circ) = 0.9861$ lagging

Example 3.11: The resistances and leakage reactances of a 30 kVA, 60 Hz, 2,400 V: 240 V distribution transformer are

$$R_1 = 0.68\ \Omega \qquad R_2 = 0.0068\ \Omega$$

$$X_{l1} = 7.8\ \Omega \qquad X_{l2} = 0.0780\ \Omega$$

Here, subscript 1 denotes the 2,400 V winding and subscript 2 denotes the 240 V winding. Each quantity is referred to its own side of the transformer.

(a) Consider the transformer to deliver its rated kVA to a load on the LV side with 230 V across the load. Find the high-side terminal voltage for a load power factor of 0.85 lagging.
(b) Find the high-side terminal voltage for a load power factor of 0.85 leading.

Solution:

(a) Using the equivalent circuit referred to the HV side, $V_L = 230\angle 0^\circ$ V

$$\text{Load current, } I_{load} = \frac{30000}{230}\angle\phi = 93.8\angle\phi \text{ A}$$

where ϕ is the p.f. angle ($\phi > 0$ for leading p.f.).
Referred to the HV side:

$$I_H = 93.8\ \angle\phi \text{ A}$$

$$V_H = V_L^{'} + Z_H I_H = 2300\angle 0^0 + (1.35 + j15.6)9.38\angle\phi$$

$$V_H = 2300 + 12.756\cos\phi - -146.328\sin\phi + j(146.328\cos\phi + 12.75\sin\phi)$$

(b) pf = 0.85 leading

$$V_H = 2233.76 + j131.1 = 2237.6\angle 3.36^0 \text{ V}$$

$$\text{pf} = 0.85 \text{ leading}$$

$$V_H = 2387.93 + j117.66 = 2390.83\ \angle 2.82^0 \text{ V}$$

Example 3.12: A single-phase load is supplied through a 35-kV feeder whose impedance is $95 + j360\ \Omega$ and a 35 kV: 2,400 V transformer whose equivalent impedance is $(0.23 + j1.27)\ \Omega$ referred to its LV side. The load is 160 kW at 0.89 leading power factor and 2,340 V.

(a) Compute the voltage at the HV terminals of the transformer.
(b) Compute the voltage at the sending end of the feeder.
(c) Compute the power and reactive power input at the sending end of the feeder.

SOLUTION:

(a) Equivalent circuit for the transformer and load:

$$P_L = |V_{load}||I_{load}|\cos\theta$$

$$|I_{load}| = \frac{160\times 10^3}{2340\times 0.89} = 76.83 \text{ A}$$

$$\theta = \cos^{-1} 0.89 = 27.13 \quad \text{leading}$$

$$I_{load} = 76.83\angle 27.13^\circ \text{ A}$$

The HV side voltage referred to the LV side:

$$V_H^{'} = V_L + Z_{eq} I_{load} = (0.23 + j1.27)(27.83\angle 27.13^\circ) + 2340$$
$$= 2311.2 + j\,94.9 \text{ V}$$

$$V_H = \frac{2.4}{35} V_H' = 33.71 + j1.384 \text{ kV}$$

$$|V_H| = 33.734 \text{ kV}$$

(b) Load current referred to the HV side:

$$I_{load}' = I_{feed} = \frac{2.4}{35} \times 76.83\angle 27.13^\circ = 5.2683\angle 27.13^\circ \text{ A}$$

$$V_{send} = Z_f I_{feed} + V_H = (95 + j360) \times 5.2683\angle 27.13^\circ + 3371 + j1384$$
$$= 33.286 + j\,3.3 \text{ kV}$$

$$|V_{send}| = 33.45 \text{ kV}$$

Example 3.13: The following data were obtained for a 20 kVA, 60 Hz, 2,400:240 V distribution transformer tested at 60 Hz:

Voltage	**Current**	**Power**
V	A	W
With high-voltage winding open-circuited		
240	1.038	122
With low-voltage terminals short-circuited		
61.3	8.33	257

Compute the efficiency at full-load current and the rated terminal voltage at 0.8 power factor.

SOLUTION:

Rated current on the H.V. side = 20 kVA/2400 = 8.33 A. Therefore, total power loss at full load current:

$$P_L = 122 + 257 = 379 \text{ W}$$

Load power at full load, $0.8 \text{ pf} = 0.8 \times 20 \text{ kW} = 16 \text{ kW}$
Therefore, input power $= 16 + 0.379 = 16.379$ kW

$$\text{Efficiency} = (16/16.379) \times 100\% = 97.7\ \%.$$

Problems

3.1 The no-load ratio required in a single-phase 50 Hz transformer is 6,600/300 V. If the maximum value of flux in the core is to be about 0.09 Wb, find the number of turns in each winding.

3.2 A 20 kVA single-phase transformer has 200 turns on the primary and 40 turns on the secondary. The primary is connected to 1,000 V, 50 Hz supply. Determine: the secondary voltage on OC; the currents flowing through the two windings on full-load; and the maximum value of flux.

3.3 A single-phase transformer is connected to a 230 V, 50 Hz supply. The net cross-sectional area of the core is 50 cm^2. The number of turns in the primary is 46 and in the secondary 80. Determine:

a. Transformation ratio;
b. Peak value of flux density in the core; and
c. EMF in the secondary winding

3.4 A 6,600/440 V single-phase 600 kVA transformer has 1,200 primary turns. Find:

a. Transformation ratio;
b. Secondary turns;
c. Voltage/turn; and
d. Secondary current when it supplies a load of 400 kW at 0.8 power factor lagging.

3.5 Find the primary and secondary turns of a 3,300/300 V, single-phase, 50 Hz, 30 kVA transformer if the flux in the core is to be about 0.06 Wb. Also determine the primary and secondary currents if the losses are to be neglected.

3.6 The voltage/turn of a single-phase transformer is 1.1 V, when the primary winding is connected to a 220 V, 50 Hz, AC supply, the secondary voltage is found to be 600 V. Find: Primary and secondary turns; and Core area if the maximum flux density is 1.2 T.

3.7 A 2,000/200 V single-phase transformer gives 0.5 A and 40 W as ammeter and wattmeter readings when supply is given to the LV winding and HV winding is kept open. Find:

a. The magnetizing component,
b. The iron loss component, and
c. The power factor of no-load current,

3.8 Find: (i) active component and reactive components of no-load current and; (ii) no-load current of a 230 V/115 V single-phase transformer if the power input on no-load to the HV winding is 70 W and power factor of no-load current is 0.25 lagging.

3.9 A single-phase transformer has 500 turns on the primary and 40 turns on the secondary winding. The mean length of the magnetic path in the core is 150 cm and the joints are equivalent to an air-gap of 0.1 mm. When a potential difference of 3,000 V is applied to the primary, maximum flux density is 1.2 T. Calculate:

a. The cross-sectional are of the core,
b. No-load secondary voltage,
c. The no-load current drawn by the primary, and
d. Power factor on no-load.

Given that AT/cm for a flux density of 1.2 T in the iron to be 5, the corresponding iron loss to be 2 W/kg at 50 Hz and density of iron as 7.8 g/cm^3.

3.10 A 230 V/115 V single-phase transformer takes a no-load current of 1.7 A at a power factor of 0.18 lagging with LV winding kept open. If the LV winding is now loaded to take a current of 13 A at 0.8 power factor lagging find the current taken by HV winding.

3.11 A transformer has a primary winding of 800 turns and a secondary winding of 200 turns. When the load current on the secondary is 80 A at 0.8 power factor lagging, the primary current is 25 A at 0.707 power factor lagging. Determine graphically or otherwise the no-load current of the transformer and its phase with respect to the voltage.

3.12 A 8:1 step down, single-phase transformer takes a no-load current of 0.6 A at a power factor of 0.8 lagging with 1.v. winding kept open. If the secondary is connected to a load taking a current of 80 A at0.8 power factor lagging; and 0.8 power factor leading find the primary current and power factor.

3.13 A 20 kVA, 2,000/200 V single-phase, 50 Hz transformer has a primary resistance of 2.5 Ω and reactance of 4.8 Ω. The secondary resistance and reactance are 0.01 Ω and 0.018 Ω respectively. Find

a. Equivalent resistance referred to primary
b. Equivalent impedance referred to primary
c. Equivalent resistance, reactance and impedance referred to secondary, and
d. Total copper loss of the transformer.

3.14 A 50 kVA, 4,400/220 V transformer has $R_1 = 3.45\,\Omega$, $R_2 = 0.009\,\Omega$, The values of reactances are $X_1 = 5.2\,\Omega$ and $X_2 = 0.015\,\Omega$. Calculate for the transformer:

a. Equivalent resistance as referred to primary
b. Equivalent resistance as referred to secondary
c. Equivalent reactance as referred to both primary and secondary,
d. Equivalent impedance as referred to both primary and secondary, and
e. Total copper loss, first using individual resistances of the two windings and secondly, using equivalent resistance as referred to each side.

3.15 A single-phase transformer has the following data:

turn ratio = 19.5:1

$R_1 = 25\,\Omega$

$X_1 = 100\,\Omega$

$R_2 = 0.06\,\Omega$

$X_2 = 0.25\,\Omega$

No load current = 1.25 A leading the flux by 30°.

3.16 The HV and LV windings of a 1,100/110 V single-phase 50 Hz transformer has resistances of 2.4 Ω and 0.02 Ω and reactance 1 Ω and 0.009 Ω respectively. The LV winding is connected to a load having an impedance of $(3 + j2)\,\Omega$. Determine:

a. Current in LV winding
b. Current in HV winding

c. The load voltage
d. Power consumed by the load.

3.17 The parameters of a 2,300/230 V, 50 Hz transformer are given below:

$R_1 = 0.286\ \Omega$ $\quad$ $R_2' = 0.319\ \Omega$

$X_1 = 0.73\ \Omega$ $\quad$ $X_2' = 0.73\ \Omega$

$R_0 = 250\ \Omega$ $\quad$ $R_0 = 1250\ \Omega$

The secondary load impedance $Z_L = 0.387 + j0.29$. Solve the exact equivalent circuit with normal voltage across the primary.

3.18 A 230 V, 3 kVA single-phase transformer has an iron loss of 100 W at 40 Hz and 70 W at 30 Hz. Find the hysteresis and eddy current losses at 50 Hz.

3.19 When a transformer is supplied at 400 V, 50 Hz, the hysteresis loss is found to be 300 W and eddy current loss is found to be 250 W. Determine the hysteresis loss and eddy current loss when the transformer is supplied at 800 V, 100 Hz.

3.20 When the primary of a transformer rated at 2,200 V, 50 Hz is supplied at 2,200 V, 50 Hz, the wattmeter gives a reading of 1,200 W on no-load. When it is supplied at 1,100 V, 25 Hz, the wattmeter gives a reading of 400 W on no-load. If the wattmeter is connected in the input circuit find the hysteresis loss and eddy current loss at normal voltage and frequency.

3.21 A 4,400 V, 50 Hz transformer has a hysteresis loss of 1,000 W, eddy current loss of 1500 W and full-load copper loss of 3,500 W. If the transformer is supplied at 6,600 V, 75 Hz, what will be the losses? Assume that the full-load current remains the same.

3.22 A 50 Hz, single-phase transformer has a turn ratio of 6. The resistances are 0.9 Ω, 0.03 Ω and reactances are 5 Ω and 0.13 Ω for HV and LV windings respectively. Find:

a. The voltage to be applied to the HV side to obtain full-load current of 200 A in the LV winding on short-circuit.
b. The power factor on short-circuit.
c. Draw the equivalent circuit and show therein all the values.

3.23 A 200/2000 V transformer is fed from a 200 V supply. The total winding resistance and leakage reactance as referred to LV side are 0.16 Ω and 7.0 Ω respectively. The representing core loss is 400 Ω and the magnetizing reactance is 231 Ω. A load of impedance $596 + j444\ \Omega$ is connected across the secondary terminals. Calculate:

a. Input current
b. The secondary terminal voltage
c. Primary power factor.

3.24 Determine the approximate equivalent circuit of a given 200/2,000 V single-phase 40 kVA transformer having the following test results:
OC test: 200 V, 6.4 A, 384 W . . . on LV side
SC test: 78 V, 20 A, 620 W . . . on HV side.

3.25 Determine the approximate circuit of a 200/400 V, 50 Hz, single-phase transformer from the following test data:
OC test (LV side): 200 V, 0.7 A, 70 W
SC test (HV side): 15 V, 10 A, 85 W
Calculate the secondary voltage when delivering 5 kW at 0.8 power factor, lagging, the primary voltage being 200 V.

3.25 A single-phase, 10 kVA, 500/250 V, 50 Hz transformer has the following constants:
Resistance: Primary 0.2 Ω, Secondary 0.5 Ω
Reactance: Primary 0.4 Ω, Secondary 0.1 Ω
Resistance or equivalent exciting circuit referred to primary, $R_0 = 1500\ \Omega$, reactance of the equivalent exciting circuit referred to primary, $X_0 = 750\ \Omega$.
What would be the readings of the instruments when the transformer is connected for the open-circuit and short-circuit tests?

3.26 Find the approximate equivalent circuit of a single-phase 400/4400 V transformer having the following test readings:
OC test (LV side): 400 V, 5.2 A, 600 W
SC test (HV side): 155 V, 50 A, 1,850 W

3.27 The following readings were taken in open-circuit and short-circuit tests on a single-phase 20 kVA, 2,000/200 V transformer:
OC test (LV side): 220 V, 2 A, 120 W
SC test (HV side): 30 V, 10 A, 140 W

Determine:
(a) Equivalent circuit referred to LV side
(b) Secondary terminal voltage as full-load at 0.8 power factor leading.

3.28 The corrected instrument reading obtained from open and short-circuit tests on 10 kVA, 450/120 V, 50 Hz transformer are:

OC Test:

$V_1 = 120$ V, $I_1 = 4.2$ A, $W_1 = 80$ W; V_1, W_1 and I_1 were read on the LV side.

SC Test:

$V_1 = 9.65$ V; $I_1 = 22.2$ A, $W_1 = 120$ W; with the LV side.

Calculate:
(a) The equivalent circuit (approximate) constants.
(b) Efficiency and voltage regulation for 0.8 lagging power.
(c) The efficiency at half full-load and 0.8 lagging power factor load.

3.29 The primary and secondary winding resistance of a 40 kVA, 6,600/250 V single-phase transformer are 10 Ω and 0.05 Ω respectively. The equivalent leakage reactance as referred to the primary winding is 35 Ω. Find the full-load regulation for load power factors of
a. Unity,
b. 0.8 lagging, and

c. 0.8 leading.

3.30 The percentage resistance and reactance of a transformer are 2.5% and 4% respectively. Find the approximate regulation on full-load at

a. Unity power factor,
b. 0.8 power factor lagging, and
c. 0.8 power factor leading.

3.31 A single-phase 100 kVA, 2,000/200 V, 50 Hz transformer has impedance drop of 10% and resistance drop of 5%.

a. Find the regulation at full-load, 0.8 power factor lagging.
b. At what power factor is the regulation zero?

3.32 The HV of a single-phase 220 kVA 4,400/220 V transformer takes a current of 30 A and power of 1,200 W at 75 V when the LV winding is short-circuited. Determine:

1. The voltage to be applied to the HV winding on full-load 0.8 power factor lagging if the secondary terminal voltage is to be kept at 220 V, and
2. % voltage regulation.

3.33 A 20 kVA, 2,200/220 V 50 Hz distribution transformer is tested for efficiency and regulation as follows;
OC test (LV side): 220 V, 4.2 A, 148 W.
SC test (HV side): 86 V, 10.5 A, 360 W.

Determine:

(i) Core loss
(ii) Equivalent resistance referred to primary.
(iii) Equivalent resistance referred to secondary.
(iv) Equivalent reactance referred to primary.
(v) Equivalent reactance referred to secondary.
(vi) Regulation of transformer at 0.8 power factor lagging current.

3.34 A 100 kVA single-phase transformer has a full-load primary current of 400 A and total resistance referred to primary is 0.006 Ω. If the iron loss amount to 500 W, find the efficiency on full-load and half-load at

a. unity power factor, and
b. 0.8 power factor.

3.35 A 50 kVA, single-phase transformer has an iron loss of 400 W and full-load copper loss of 800 W.

a. Find the load at which maximum efficiency occurs and the value of maximum efficiency at unity power factor.
b. If the maximum efficiency occurs at 80% full load, find the new core loss and full-load copper loss assuming that total full-load loss is constant.

3.36 A 1,100/220 V transformer has a primary resistance of 0.25 Ω and secondary resistance of 0.03 Ω. If iron loss amounts to 250 W, determine the secondary current at which maximum efficiency occurs and find the maximum efficiency at 0.8 power factor.

3.37 A single-phase 250 kVA transformer has an efficiency of 96% on full-load at 0.8 power factor and on half full-load at 0.8 power factor. Find:

a. Iron loss, and
b. full-load copper loss.

3.38 The efficiency of a 300 kVA, single-phase transformer is 97.8% when delivering full-load at 0.8 power factor lagging and 98.4% when delivering half full-load at unity power factor. Determine the efficiency at 80% of full-load at 0.8 power factor lagging.

3.39 A 12 kVA, 400/200 V single-phase 50 Hz transformer has maximum efficiency of 95% at 85% of full-load at unity power factor. Determine the efficiency at full-load at 0.8 power factor lagging.

3.40 A 50 kVA, 2,000/250 V single-phase transformer has resistance of 1.1 Ω and 0.015 Ω and reactance of 4.4 Ω and 0.06 Ω for the HV and LV windings, respectively. It has a maximum efficiency of 96.05% at 80% of full-load at unity p.f. The magnetizing current for the HV side at 2,000 V is 1.25 A. Find the readings of suitable instruments for OC test and SC tests, supply being given to HV side in both cases.

3.41 The maximum efficiency of a single-phase 240 kVA, 2,000/250 V transformer occurs at 70% of full-load and is equal to 98% of 0.8 p.f. lagging. Determine the efficiency and regulation on full-load at 0.8 p.f. lagging if the impedance of the transformer is 8%.

3.42 The maximum efficiency of a single-phase 240 kVA, 2,000/250 V transformer occurs at 70% of full-load and is equal to 98% at 0.8 p.f. lagging. Determine the efficiency and regulation on full-load at 0.8 p.f. lagging if the impedance of the transformer is 8%.

3.43 A 5 kVA, 230/115 V transformer takes 1.2 A and 75 W when 230 V is applied to HV winding and LV winding is kept open. It takes 21.75 A and 150 W when the LV winding is short-circuited, and 17.4 V is applied to the HV winding. Find:

a. The no-load current as a fraction of full-load input current at 0.8 p.f. lagging;
b. Percentage regulation on full-load input current at 0.8 p.f. lagging; and
c. The load at which maximum efficiency occurs and the maximum efficiency at unity p.f.

3.44 The efficiency of a 20 kVA, 2,000/200 V transformer is 96.8% at full-load at unity p.f. and 96% at full-load at unity p.f. and 96% at 60% of 0.8 p.f. lagging; and 0.8 p.f. leading if the impedance is 7%.

3.45 A 100 kVA, 2,000/200 V single-phase transformer takes a current of 50 A and 2,400 W at 100 V when the LV winding is short-circuited. Determine the LV and percentage regulation when delivering full-load current at 0.8 p.f. leading, the supply voltage being 2,000 V.

3.46 A single-phase, 25 kVA transformer has an iron loss of 240 W and full-load copper loss of 600 W. Find the load at which maximum efficiency occurs and maximum efficiency at 0.8 p.f.

If the maximum efficiency occurs at 70% of full-load, find the core loss and copper loss assuming the total loss to be the same as in the previous case.

4 Synchronous Generators

4.1 INTRODUCTION

A synchronous machine is an AC rotating machine whose speed under steady state condition is proportional to the frequency of the current in its armature. The magnetic field created by the armature currents rotates at the same speed as that created by the field current on the rotor, which is rotating at the synchronous speed, and a steady torque results. Synchronous machines are commonly used as generators especially for large power systems, such as turbine generators and hydroelectric generators in the grid power supply. Because the rotor speed is proportional to the frequency of excitation, synchronous motors can be used in situations where constant speed drive is required.

4.2 SYNCHRONOUS MACHINE STRUCTURE

Synchronous machine is consisting of two parts, one is stator, and another is rotor:

The stator or armature is an iron ring, formed of laminations of silicon steel with slots in periphery to contains armature conductors. These slots may be open, semi-closed and closed according to speed and size of machine. Open slots are most commonly used because the coil can be freely wound and insulated properly. These slots provide the facility of removal and replacement of defective coils. The semi-closed slots are used to provide better performance over open slots. The totally closed slots are rarely used.

A synchronous generator is an electromechanical device which converts mechanical energy (usually provided by steam, water or gas turbine as the "prime-mover") into electrical energy in the form of 3-phase (usually) AC quantities. It works on the principle of Faraday's Law of Electromagnetic Induction. Synchronous Generators are known as Alternator. The term "Synchronous Generator" usually refers to a machine in a Power Station connected to a large, interconnected power system.

Electromechanical energy conversion takes place whenever a change in flux is associated with mechanical motion. EMF is generated in a coil when there is a relative movement between the coil and the magnetic field. Alternating EMF is generated if the change in flux-linkage of the coil is cyclic. Since electromechanical energy conversion requires relative motion between the field and armature winding, either of these could be placed on the stator or rotor. Because of practical convenience, field windings are normally placed on the rotor and the stator serves as the seats of induced EMF (i.e., the armature winding will be on stator) in almost all Synchronous machines. There are two types of field construction used on synchronous machines: salient and cylindrical rotor or non-salient pole as shown in Figure 4.1. The rotor is rotated at the synchronous speed by a prime mover such as a steam turbine. The rotor

DOI: 10.1201/ 9781003795841-4

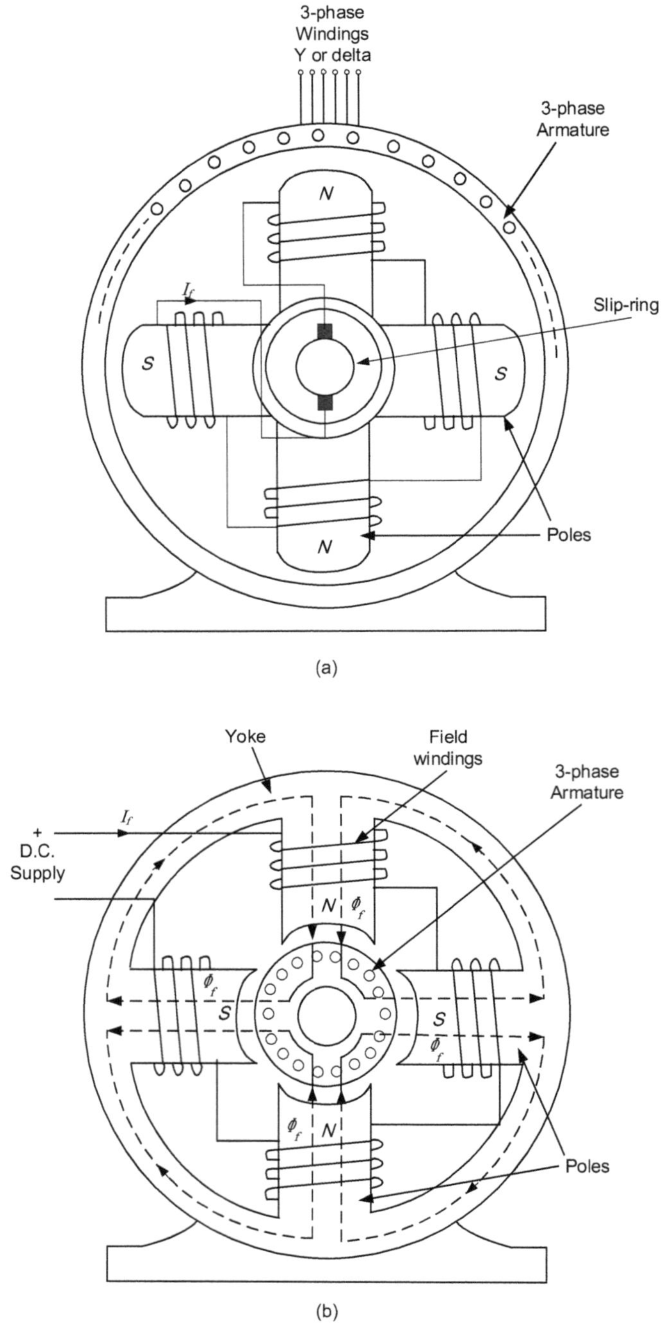

FIGURE 4.1 Schematic illustration of synchronous machines of (a) salient rotor structures and (b) round or cylindrical rotor.

has as many poles as the stator, and the rotor winding carries DC current so as to produce constant flux per pole.

The salient-pole rotor (Figure 4.1a) is used in low- and medium-speed generators because the windage loss is small at these speeds. It consists of an even set of outward projecting laminated poles. Each pole is dove-tailed so that it fits into a wedge-shaped recess or is bolted onto a magnetic wheel called the spider. The field winding is placed around each pole, as indicated in Figure 4.1a. These salient-pole machines are designed with stator armature of large diameter and short axial length. The purpose of these special dimensions for salient-pole machines is to build a machine where proper ventilation and dissipation of heat generated inside it are simple and effective.

The cylindrical rotor (Figure 4.1b) is designed to stand high speed drives since there is no shear force due to the absence of saliency in the poles. The prime movers used with this type are steam turbines or gas turbines. The cylindrical rotor is employed in a 2- or 4-pole since steam turbines are most efficient at high speeds. It is now obvious that a large diameter is prohibited for this type of machine because of the increased centrifugal force at these high speeds. But, the small diameter necessitates fewer armature conductors, so that in order to obtain the required generated voltage a greater length of conductor must be used. In other words, the axial length is increased to permit the use of longer armature conductors. The armature is made of a smooth solid forged steel cylinder with a number of slots on its outer periphery. These slots are designed to accommodate the field coils, as shown in Figure 4.1b. The cylindrical construction offers the following benefits:

1. It results in a quiet operation at high speed;
2. It provides better balance than the salient-pole rotor;
3. It reduces the windage loss.

4.3 ROTATING FIELD

The magnetic field required for the generation of AC voltage is provided by rotating magnetic field similar to DC generator. If the 3-phase winding is powered by a 3-phase system, currents I_1, I_2, and I_3 will create at the same time their own magnetic fields H_1, H_2 and H_3. These fields are spaced between them by a 120° angle as well. Besides that, as they are proportional to the respective currents, they will be de-phased in time, also between them by a 120° angle. The resulting H field, at each point, will be equal to the graphic sum of the 3 magnetic fields H_1, H_2, and H_3 on that point as shown in Figure 4.2.

Another field system is placed on a rotating shaft, which rotates within the armature conductors or stator. The field system contains electromagnets which are excited by pilot or main exciters. Generally main exciters are used but for very large machines the pilot exciter is also used. These exciters are DC generators.

The various advantages of rotating field can be stated as,

1. In applications where AC is prevalent, the generated AC voltage may range from 11 kV to 33 kV, which is induced in the armature. For a stationary

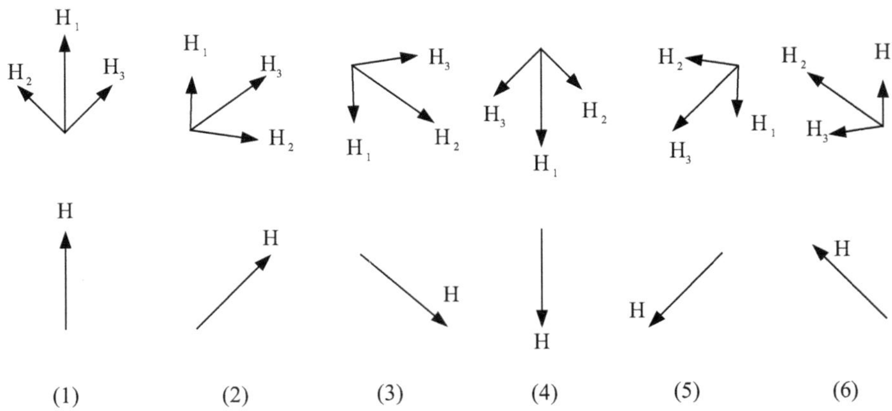

FIGURE 4.2 The rotating magnetic field of 3-phase synchronous machine.

armature, ample space can be provided to accommodate a large number of conductors along with their insulations.

2. It is advisable to shield high voltage windings from centrifugal forces resulting from rotation. Therefore, the high voltage armature is typically kept stationary, mitigating the effects of mechanical and electrical stresses.
3. Collecting large currents at very high voltages is more feasible from a stationary component than from a slip ring and brush assembly. The relatively low voltage required for the field (110 V to 220 V DC) can be easily supplied using a rotating slip ring and brush assembly.
4. By maintaining a rotating field which operates at low voltage, issues such as sparking at the slip rings can be minimized, with the high voltage armature remaining stationary.
5. Since the field side operates at a low voltage, less insulation is required, resulting in very low inertia in the field system. It is generally more efficient to rotate a system with low inertia than one with high inertia, as less effort is needed to rotate a low inertia system.
6. A rotating field simplifies the overall construction of the system. With a straightforward, sturdy mechanical structure and low inertia in the rotor, it can operate at higher speeds, thereby producing more output from a given size of alternator.
7. If the field is rotating, exciting it with an external DC supply requires only two slip rings – one for each of the positive and negative terminals. In contrast, a 3-phase rotating armature would need at least three slip rings, which are challenging to insulate at high voltage levels.
8. Ventilation arrangements for the high voltage side can be enhanced by keeping it stationary.

Due to all these reasons most of the alternators in practice use rotating field type of arrangement. For small voltage rating alternators rotating armature arrangement may be used.

4.4 EMF EQUATION

The synchronous machine windings may be classified as concentric windings and distributed windings. In a concentric winding, every coils have the same axis and hence, the pitches of the coils forming a coil group are different. Concentric windings are always single layer.

The stators (armatures) of most synchronous generators feature three separate and distinct windings that produce 3-phase power. Each winding corresponds to one phase of the 3-phase generator. These windings are identical in shape and configuration but are spatially offset from each other by exactly 120 degrees electrical. This specific arrangement ensures that the electromotive forces (EMFs) induced in these windings are exactly 120° electrical out of phase in terms of timing.

The 3-phase windings can be configured in either a star (Y) or delta (Δ) connection. When configured in a Y connection, the windings are internally connected, and the neutral point is extended to allow proper grounding. This grounding is crucial for safe operation, as illustrated in Figure 4.3.

The armature of an asynchronous generator is often constructed using a double-layer winding. This type of winding arrangement requires that there be as many identical coils as there are slots in the stator. In this setup, one side of each coil is positioned in the bottom half of a slot, while the other side occupies the top half of another slot. To achieve this, coils are pre-wound on forms before being inserted into the slots. The number of coils per phase or the number of slots per phase in the case of a double-layer winding must be an integer. Additionally, because the coils need to be evenly distributed across the poles, the number of coils (or slots) per pole per phase must also be an integer. Therefore, if $2P$ represents the total number of slots on the armature, $2P$ the total number of poles, and m the number of phases, the formula to determine the number of coils per pole per phase is as follows:

$$n = \mathrm{S}/(2 \times P \times \mathrm{m})$$

where n must be an integer. The number of coils per pole per phase, n, is usually referred to as a phase group or phase belt.

Let $2P$ be the number of poles, Φ the flux per pole in weber, N the speed of the alternator in rpm, T the total number of turns per phase, and f the frequency of generated EMF in Hz, then,

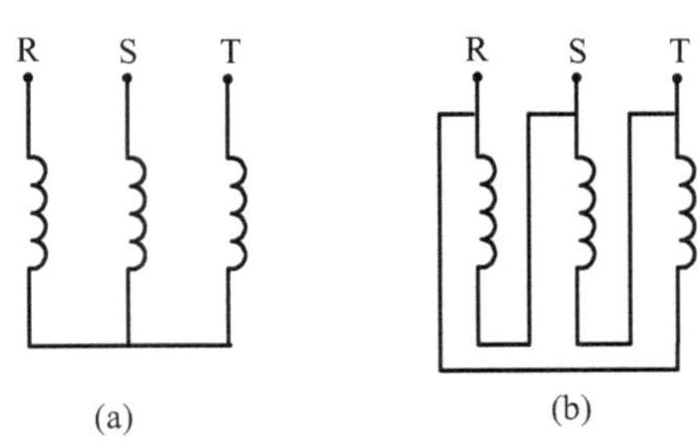

FIGURE 4.3 Y and Δ connection.

$$\text{Flux cut by each conductor per revolution} = 2P \times \Phi$$

$$\text{Flux cut by each conductor per revolution} = 2P \times \Phi \times N/60$$

$$\text{Thus, average EMF generated in each conductor} = 2P \times \Phi \times N/60 \text{ Vs}$$

Since, one turn consists of two active conductors,

$$\text{The average EMF generated per turn} = 2 \times 2P \times \Phi \times N/60 \text{ Vs}$$

$$\text{But } f = P\,N/60 \text{ Hz}$$

Thus, average EMF per phase $= 4f\Phi T$ Vs

RMS value of EMF generated per phase,

$$E = 1.11 \times 4 \times f \times \Phi \times T \text{ Vs}$$
$$= 4.44 \times f \times \Phi \times T \text{ Vs}$$

In the derivation of the EMF equation, the distribution of the winding in various slots and the chording the winding have not been considered. When proper consideration is taken into account for these two effects, the rms value of the EMF generated per phase is given by,

$$E = 1.11 \times 4 f\Phi\, T\, k_d\, k_c \text{ Vs}$$

where k_d the distribution factor and k_c the coil span or pitch factor.

However, k_c and k_d can be represented by a single term k_w, known as the winding factor

$$\text{Hence,} \qquad E = 1.11 \times 4f\,\Phi\, Tk_d\, k_c \text{ Vs} \qquad (4.1)$$

In concentrated winding, all the coil sides will be placed in one slot under a pole. So induced EMF in all the coils will achieve maxima and minima at the same time, that is, all of them will be in phase. Hence resultant EMF. after connecting coils in series will be algebraic sum of all the EMFs as all are in phase as shown in Figure 4.4. Concentrated winding is also rare in practice. Attempt is made to use all the slots available under a pole for the winding which makes the nature of the induced EMF more sinusoidal as shown in Figure 4.5.

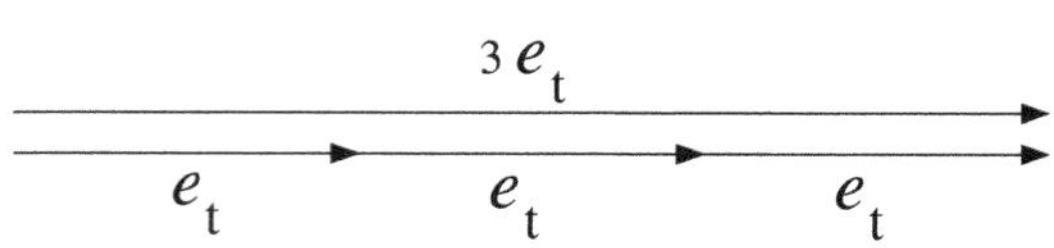

FIGURE 4.4 EMF of full-pitch coil.

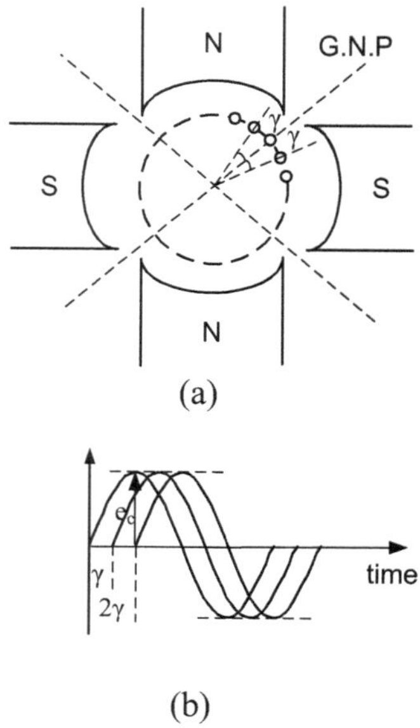

FIGURE 4.5 Generated EMF in distributed windings.

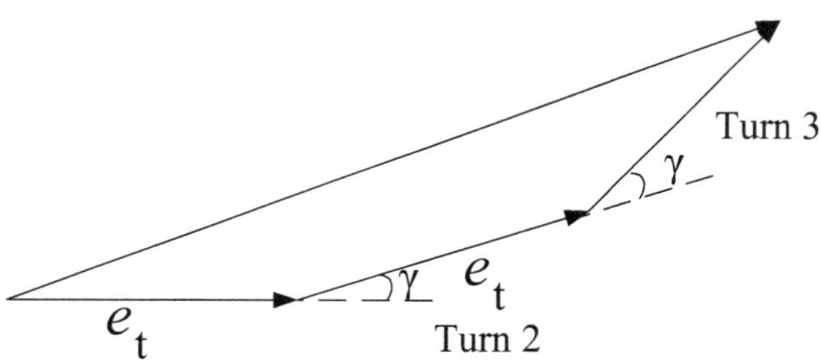

FIGURE 4.6 EMF of chorded-pitch coil.

With a distributed winding, the EMF induced in the various coils constituting a group of coils are not in phase and so the total EMF is the vector sum of all such EMF's as shown in Figure 4.6. Hence, its value is less than that due to a concentrated winding. A factor K_d (which is less than unity) is therefore, introduced in the expression of EMF.

Consider 18 slots, 2-pole alternator. So slots per pole

$$m = \text{slots per pole per phase} = 3$$

$$\gamma = 360\ \text{P/S} = 360/18 = 20^{\circ}$$

Assume that the number of conductors per phase per pole is 3 and the conductors are connected in lap connection as shown in Figure 4.7.

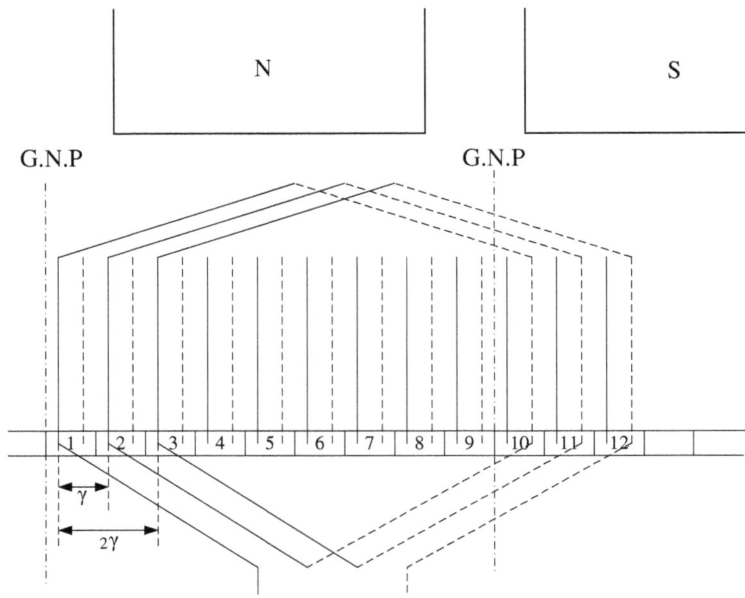

FIGURE 4.7 Distributed windings.

In general let there be "n" slots per pole and "q" slots per pole per phase. So there will be "q" turns distributed under a pole per phase, connected in series. Let E be the induced EMF per turn. Then all the "q" EMFs induced in the turns will have successive phase angle difference of $\gamma = 180^\circ/\text{n}$. While finding out the phasor sum of all of them, phasor diagram will approach a shape of a "m" equal sided polygon circumscribed by a semicircle of radius "R."

This is shown in the Figure 4.8 ab, bc, cd, etc., represents EMF per turn. All the ends joined at "o" which is center of the circumscribing semicircle of radius "R."

Angle subtended by each phasor at the origin "O" is γ. This can be proved as below. All the triangles oab, obc are similar and isosceles, as $ab = bc = cd = E$.

$$oab = \quad oba = \quad obc = x$$

$$\text{let} \quad oab = \quad obc = y$$

$$\text{Now in } \Delta\, oab, \qquad 2x + y = 180$$

$$\text{while} \quad oba + \quad obc + \gamma = 180^\circ \tag{4.2}$$

$$\text{i.e. } 2x + \gamma = 180^\circ \tag{4.3}$$

Comparing (4.2) and (4.3), $y = \gamma$

$$\text{So} \qquad aob = \quad boc = \quad cod = \gamma$$

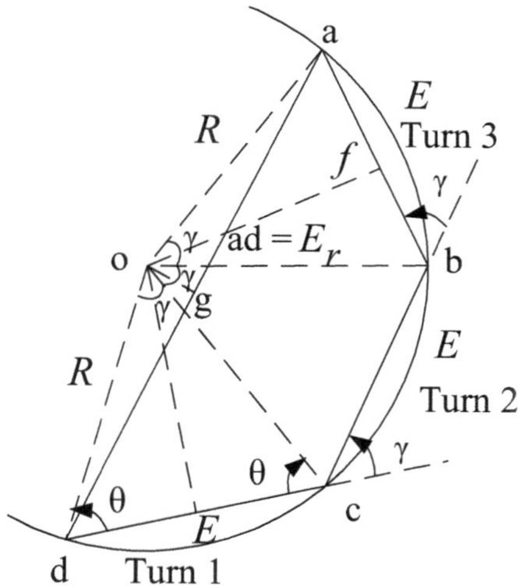

FIGURE 4.8 Phasor sum of "q" EMFs

If "d" is the last point of the last phasor,

$$aod = \mathrm{m} \times \mathrm{q} = \mathrm{q}\,\gamma$$

and ad = E_R = Resultant of all the EMF.

in Δoab, let of be the perpendicular drawn on ab bisecting angle at apex "o" as $\gamma/2$.

$$l(ab) = E \quad \text{and} \quad l(af) = E/2$$

$$\text{and} \;\; l(oa) = R.$$

$$\therefore \;\; \sin(\gamma/2) = af/oa = (E/2)/R$$

$$\therefore \;\; E = 2R\sin(\gamma/2) \qquad (4.4)$$

Now consider Δ oad and og is the perpendicular drawn from "o" on its base bisecting *oad*.

$$\therefore \quad aog = \quad god = (m\,\gamma)/2$$

$$\therefore \;\; l(ad) = E$$

$$\therefore \;\; l(ag) = E/2$$

$$= 2\,(R)\sin\,(m\,\gamma/2)$$

Arithmetic sum of EMFs = 2 (R)q × $sin\left(\gamma/2\right)$

$$k_d = \frac{Phasor\ sum}{Arithmetic\ sum} = \frac{2(R)\sin\frac{q\gamma}{2}}{2(R)q\sin\frac{\gamma}{2}}$$

$$k_d = \frac{\sin\frac{q\gamma}{2}}{q\sin\frac{\gamma}{2}} \tag{4.5}$$

To reduce weight of copper, the stator winding is constructed with a coil width less than the pole-pitch. In practice short pitch coils are preferred. So coil is form by connecting one coil side to another which is less than one pole pitch away. So actual span is less than 180°. The coil is generally shorted by one or two slots. The angle by which coil are short pitched is called angle of short pitch denoted as "ε."

Now let E be the induced EMF in each coil side. If coil is full pitch coil, the induced EMF in each coil side help each other. Coil connections are such that both will try to set up a current in the same direction in the external circuit. Hence the resultant EMF across a coil will be algebraic sum of the two as shown in Figure 4.9a.

$$\therefore\quad E_R = E + E = 2E \qquad \text{.......... for full pitch}$$

Now the coil is short pitched by angle "ε," the two EMFs in two coil sides no longer remains in phase from external circuit point of view. Hence the resultant EMF also no longer remains algebraic sum of the two but becomes a phasor sum of the two as shown in the Figure 4.7b. Obviously E_R in such a case will be less than what it is in case of full pitch coil.

From the geometry of the Figure 4.9b, we can write, ac is perpendicular drawn on ob bisecting ob.

$$\therefore\quad l(oc) = l(cb) = E_R/2$$

$$\text{and}\quad boa = \varepsilon/2$$

$$\therefore\quad \cos(\varepsilon/2) = oc/oa = E_R/2E$$

$$\therefore\quad E_R = 2\,E\cos(\varepsilon/2) \qquad \text{.............. For short pitch}$$

This is the resultant EMF in case of a short pitch coil which depends on the angle of short pitch "ε." Now the factor by which, induced EMF gets reduced due to short pitching is called pitch factor or coil span factor denoted by K_c.

It is defined as the ratio of resultant EMF when coil is short pitch to the resultant EMF when coil is full pitched. It is always less than one.

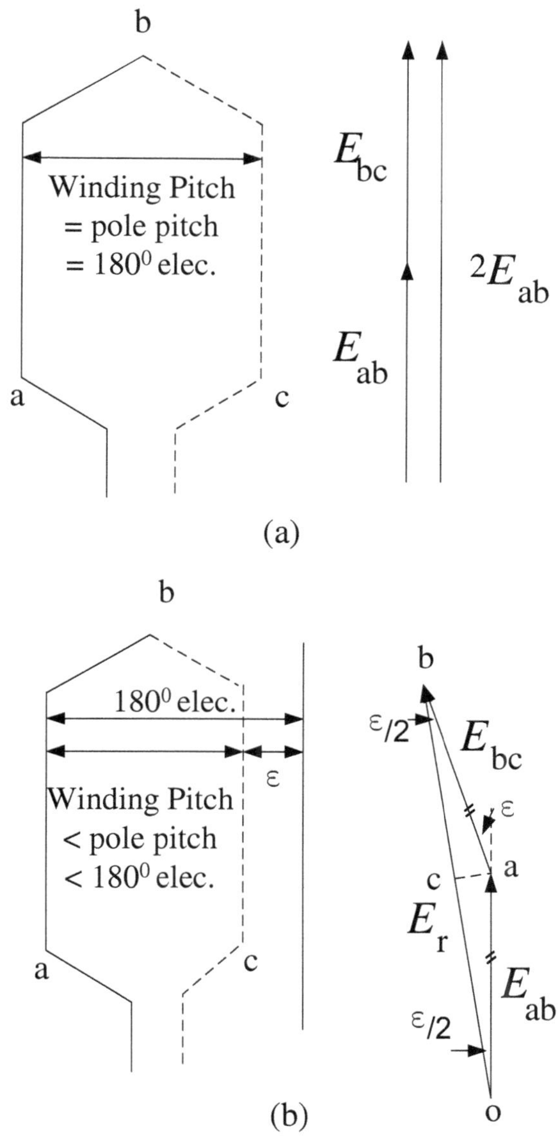

FIGURE 4.9 Full and chorded pitch coil.

$$K_c = \frac{Er \textit{ when coil is short pitched}}{Er \textit{ when coil is full pitched}} = \frac{2\,E\cos\frac{\varepsilon}{2}}{2\,E}$$

$$K_c = \cos\frac{\varepsilon}{2} \tag{4.6}$$

Final EMF expression becomes

$$E_{ph} = 4.44\, K_d\, K_c\, f\, N_{ph}\, \Phi \qquad (4.7)$$

Example 4.1: Calculate the pitch factors for the following windings:

(a) 36 slots, 4 poles, span 1 to 8
(b) 72 slots, 4 poles, span 1 to 15
(c) 96 slots, with coils having 5/6 fractional pitch

Solution:

(a) Full-pitch coil span $= \dfrac{\textit{number of slots}}{\textit{number of poles}} = \dfrac{36}{4} = 9$ slots

The coil span, p = 7 slots

$$= 7 \times (180/9) = 140 \text{ electrical degree}$$

$$\therefore \text{the pitch factor } K_c = \cos\frac{(180-140)}{2} = \cos\frac{40}{2} = 0.94$$

(b) Full-pitch coil span $= \dfrac{\textit{number of slots}}{\textit{number of poles}} = \dfrac{72}{4} = 18$ slots

The coil span, p = 14 slots

$$= 14 \times (180/18) = 140 \text{ electrical degree}$$

$$\therefore \text{the pitch factor } K_c = \cos\frac{(180-140)}{2} = \cos\frac{40}{2} = 0.94$$

(c) The coil span, $p = \dfrac{5}{6} \times 180 = 150$ electrical degree

$$\therefore \text{the pitch factor } K_c = \cos\frac{(180-150)}{2} = \cos\frac{30}{2} = 0.966$$

Example 4.2: A 3-phase, 16-pole synchronous generator has a star-connected winding with 144 slots and 10 conductors per slot. The flux per pole is 0.04 Wb and the speed is 375 rpm. Find the frequency and phase EMF and line EMF The total turns/phase may be assumed to be series connected.

Solution:

$$N = \frac{60f}{p} \Rightarrow f = \frac{PN}{60}$$

$$\therefore \quad f = \frac{8 \times 375}{60} = 50 \text{ HZ}$$

$$Z = No.\ of\ slots \times No.\ of\ Conductors\ per\ slot$$

$$Z = 144 \times 10 = 1440\ \Omega$$

$$Total\ no.\ of\ turns = \frac{1440}{2} = 720$$

$$Total\ no\ of\ turns\ perphase,\ N_{ph} = \frac{720}{3} = 240$$

$$\text{Phase EMF, } E_{ph} = 4.44\ K_d\ K_c\ f\ N_{ph}\ \Phi \qquad \text{(V)}$$

$$f = 50 \text{ Hz}, \qquad N_{ph} = 240, \qquad \Phi = 0.04 \text{ Wb},$$

$$K_c = 1, \quad K_d = \frac{\sin\left(\dfrac{q\gamma}{2}\right)}{q \sin\left(\dfrac{\gamma}{2}\right)}$$

$$m = \frac{\text{no. of slots}}{\text{no. of poles} \times \text{no. of phases}} = \frac{144}{16 \times 3}$$

$$\gamma = slot\ angle = \frac{180^\circ}{144/16} = 20^\circ$$

$$K_d = \frac{sin\left(\dfrac{3 \times 20^\circ}{2}\right)}{3sin\left(\dfrac{20^\circ}{2}\right)} = 0.96$$

$$E_{ph} = 4.44\ (0.96)(1)(50)(240)(0.04) = 2046 \qquad \text{V}$$

$$E_L = \sqrt{3}\ E_{ph} = \sqrt{3} \times 2046 = 3543.7 \qquad \text{V}$$

Example 4.3: A 3-phase, 16-pole alternator has a star-connected winding with 144 slots and 10 conductors per slot. The chording is by one slot. The flux per pole is 0.03 Wb. sinusoidally distributed and the speed is 375 rpm. Find the frequency, phase voltage and line voltage.

SOLUTION:

Given 2P = 16, $\Phi = 0.03$, N = 375 rpm

$$N = \frac{60f}{P} \Rightarrow f = \frac{PN}{60}$$

$$\therefore \quad f = \frac{8 \cdot 375}{60} = 50 \quad \text{Hz}$$

$$Z = No.\ of\ slots \times No.\ of\ Conductors\ per\ slot$$

$$Z = 144 \times 10 = 1440\ \Omega$$

$$Total\ no.\ of\ turns = \frac{1440}{2} = 720$$

$$Total\ no.\ of\ turns\ perphase,\quad N_{ph} = \frac{720}{3} = 240$$

$$\text{Phase EMF, } E_{ph} = 4.44\ K_d\ K_c\ f\ N_{ph}\ \Phi \quad \text{V}$$

$$F = 50\ \text{Hz}, \qquad N_{ph} = 240, \qquad \Phi = 0.04\ \text{Wb},$$

$$K_c = 1, \quad K_d = \frac{\sin\left(\frac{q\gamma}{2}\right)}{q\sin\left(\frac{\gamma}{2}\right)}$$

$$m = \frac{no.\ of\ slots}{no.\ of\ poles \times no.\ of\ phases} = \frac{144}{16 \times 3}$$

$$\gamma = slot\ angle = \frac{180^\circ}{9} = 20^\circ$$

$$K_d = \frac{sin\left(\frac{3 \times 20^\circ}{2}\right)}{3sin\left(\frac{20^\circ}{2}\right)} = 0.96$$

The coil is short chorded by 1 slot

$$\text{Pitch factor,}\quad K_c = \cos\frac{\varepsilon}{2} = \cos\frac{20}{2} = 0.9848$$

$$E_{ph} = 4.44\ (0.9597)\ (0.9848)(50)(240)(0.03) = 1503.5\ \text{V}$$

$$E_L = \sqrt{3}\ E_{ph} = \sqrt{3} \times 2046 = 2604.14\ \text{V}$$

Example 4.4: A 72 slots, 3-phase, whole-coil stator armature has coil span of slots, 10 turns/coils. The armature is Y-connected. The 6-pole rotor has a flux of 5.2×10^6 lines per pole and is driven at a speed of 1,200 rpm. Calculate: the effective phase voltage and line voltage.

SOLUTION:

$$\text{The full pitch} = \frac{72}{6} = 12\ \text{slots}$$

$$\text{the coil span, p} = 10(180/12) = 150\ \text{electrical degree}$$

$$\text{the pitch factor,}\ \ K_c = \cos\frac{180-150}{2} = \cos\frac{30}{2} = 0.966$$

$$\text{the number of slots per pole per phase q} = \frac{72}{6 \times 3} = 4\ \frac{72}{6 \times 3} = 4\ \text{slots}$$

$$\gamma = \mathit{Slot\ angle} = \frac{180°}{12} = 15°\ \text{electrical degree}$$

$$\text{the distribution factor}\ K_d = \frac{\sin\left(\dfrac{4 \times 15°}{2}\right)}{4\sin\left(\dfrac{15°}{2}\right)} = 0.958$$

$$\mathit{Total\ no.\ of\ turns\ perphase,}\ \ N_{ph} = \frac{72 \times 1 \times 10}{3} = 240\ \text{turns}$$

$$N = \frac{60f}{P} \Rightarrow f = \frac{PN}{60}$$

$$\therefore \qquad f = \frac{3 \cdot 1200}{60} = 60\ \text{Hz}$$

$$\text{Phase EMF,}\ E_{ph} = 4.44\ K_d\ K_c\ f\ N_{ph}\ \Phi\ \ \text{V}$$

$$E_{ph} = 4.44(0.958)(0.966)(60)(240)(5.2 \times 10^6)10^{-8} = 3080\quad \text{V}$$

$$E_L = \sqrt{3}\, E_{ph} = \sqrt{3} \times 3080 = 5330\ \ \text{V}$$

ADVANTAGES OF USING CHORDED WINDINGS

1. The waveform of the EMF induced is improved as the EMF is not sinusoidal, because the flux density under pole is not sinusoidal.
2. The saving in using cupper due to less length.
3. The overhang leakage reactance is reduced in case of chorded windings due to the length of overhang.
4. The mechanical strength of the chorded coil is better than that for full-pitch coil.

ADVANTAGES OF USING DISTRIBUTED WINDINGS

1. Distributing the windings in more slots will reduce the slot size which reduce the overheat of centered conductors.
2. Distributing the windings in more slots per pole will improve the wave shape of the EMF induced in conductors.

4.5 HARMONIC EMF

The flux density distribution around the air gap in all well designed alternators is symmetrical with respect to abscissa and also to polar axis. Thus, it can be expressed with the help of a Fourier series which do not contain any even harmonics.

So flux density at any angle from the interpolar axis is given by,

$$B = B_{m1} \sin\theta + B_{m3} \sin 3\theta + \ldots + B_{mx} \sin n\theta + \ldots$$

where x = Order of harmonic component which is odd

B_{m1}= Amplitude of fundamental component of flux density

B_{m3}= Amplitude of third harmonic component of flux density

B_{mn}= Amplitude of nth (odd) harmonic component of flux density

The EMF generated in a conductor on the armature of a rotating machine is given by,

$$e_c = B\,l\,v$$

Substituting value of B,

$$e_c = (B_{m1} \sin\theta + B_{m3} \sin 3\theta + \ldots + B_{mx} \sin n\theta + \ldots\ldots)l\,v$$

l = Active length of conductor in meter

D = Diameter of the armature at the air gap

v = Linear velocity $= \pi D n_s$

where n_s = Synchronous speed in rps.

$$\text{Now} \quad N_s = 60\, f/P$$

$$\therefore \quad n_s = 60f/60P = f/P$$

$$\therefore \quad v = \pi\ Dd\left(f/P\right)$$

Substituting in the expression for e_c

$$e_c = \left[B_{m1}\cdot\sin\theta + B_{m3}\cdot\sin 3\theta + \ldots\ldots + B_{mn}\cdot\sin n\theta\right] l\cdot\frac{\pi D}{P} f$$

$$e_c = \left[B_{m1}\cdot\frac{\pi Dl}{P}\cdot f\sin\theta + B_{m3}\cdot\frac{\pi Dl}{P}\cdot f\sin 3\theta + \ldots\ldots B_{mn}\cdot\frac{\pi Dl}{P}\cdot f\sin n\theta\right] \quad (4.8)$$

Area of each fundamental pole, $A_1 = (\pi\ D\ l)/2\mathrm{P}$

$$\therefore \quad e_c = \left(B_{m1}\ A_1\ \ 2f\ \sin\theta + B_{m3}\ A_1\ \ 2f\ \ \sin 3\theta + \ldots. + B_{mx}A_1\ 2f\ \sin x\theta\ \right)$$

Area of xth harmonic pole, $\mathrm{A_n} = (\pi\ \mathrm{D}\ l)/\left(2\mathrm{nP}\right) = \mathrm{A_1}/\mathrm{x}$

This is because, there are 2nP poles for the nth order harmonic

$$\therefore \quad e_c = 2f\left(B_{m1}\ A_1\ \sin\theta + B_{m3}\ 3A_3\ \sin 3\theta + \ldots. + B_{mx}A_{x\ \sin x}\theta\right)$$

Considering the windings to short pitch and distributed, the EMF of a fundamental frequency is given by,

$$E_{ph1} = 4.44\ K_{d1}\ K_{c1}\ f\ N_{ph}\ \Phi_1, \quad (4.9)$$

While the phase EMF of order harmonic component of frequency is given by,

$$E_{ph3} = 4.44\ K_{d3}\ K_{c3}\ 3f\ N_{ph}\ \Phi_3 \quad (4.10)$$

$$E_{phn} = 4.44\ K_{dn}\ K_{cn}\ nf\ N_{ph}\ \Phi_n \quad (4.11)$$

The total phase EMF is given by,

$$E_{ph} = \sqrt{\left(E_{ph1}\right)^2 + \left(E_{ph3}\right)^2 + \left(E_{ph5}\right)^2 + \ldots\ldots + \left(E_{phn}\right)^2} \quad (4.12)$$

we know that,

ε = Angle of short pitch for fundamental flux wave

The changes for various harmonic component of flux as,

3 ε = For third harmonic component

5 ε = For fifth harmonic component

.

.

.

n ε = For n[th] harmonic component

Hence the pitch factor is expressed as,

$$k_c = \cos\frac{n\,\varepsilon}{2}$$

where n = Order of harmonic component

Similar to the pitch factor, the distribution factor is also different for various harmonic components.

The general expression to obtain distribution factor is,

$$k_d = \frac{\sin\frac{q\gamma}{2}}{q\sin\frac{\gamma}{2}}$$

where x = order of harmonic component

$$\text{For } x = 1, \qquad k_{d1} = \frac{\sin\frac{q\gamma}{2}}{q\sin\frac{\gamma}{2}}$$

$$\text{For } x = 3, \qquad k_{d3} = \frac{\sin\frac{3q\gamma}{2}}{q\sin\frac{3\gamma}{2}}$$

For star connected, the line or terminal induced EMF is √3 times the total phase EMF, but it should be noted that with star connection, the third harmonic voltages do not appear across line terminal though present in phase voltage.

$$E_{line} = \sqrt{\left(E_{ph1}\right)^2 + \left(E_{ph5}\right)^2} \tag{4.13}$$

In delta connection also, 3rd, 9th, 15th. . . harmonic voltages do not appear at the line terminals. Taking ratio of fundamental frequency EMF and n^{th} order harmonic frequency EMF we can write,

$$\frac{E_{1ph}}{E_{nph}} = \frac{K_{c1}K_{d1}B_1}{K_{cn}K_{dn}B_n} = \frac{K_{c1}K_{d1}}{K_{cn}K_{dn}}\cdot\frac{B_1}{B_n}$$

$$\frac{E_{1ph}}{E_{nph}} == \frac{K_{w1}}{K_{wn}}\cdot\frac{B_1}{B_n} \tag{4.14}$$

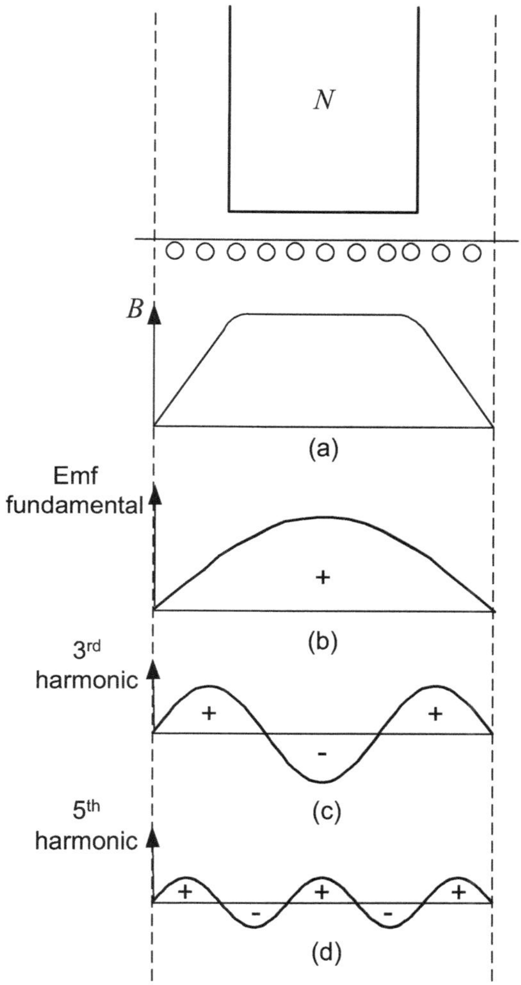

FIGURE 4.10 EMF harmonics.

Example 4.5: A 3-phase, star-connected, 50 Hz, 6-pole alternator has a fundamental flux per pole of 0.015 Wb. The stator has 90 slots and 4 conductors per slot and each stator coil span 12 slots.

If the flux in the air gap contains 30% third harmonics and a 20% fifth harmonic with respect to the fundamental, determine the rms values of per phase and line induced EMFs.

SOLUTION:

$$\text{The full pitch} = \frac{90}{6} = 15 \text{ slots}$$

the coil span, $p = 12 \times (180/15) = 144$ electrical degree

the pitch factor for fundamental,

$$K_{c1} = \cos\frac{180-144}{2} = \cos\frac{36}{2} = 0.951$$

the pitch factor for third harmonic,

$$K_{c3} = \cos\frac{3*36}{2} = \cos\frac{108}{2} = -0.588$$

the pitch factor for fifth harmonic,

$$K_{c5} = \cos\frac{5*36}{2} = \cos\frac{180}{2} = 0$$

the number of slots per pole per phase $q = \dfrac{90}{5\times 3} = 5$ slots

$$\gamma = Slot\ angle = \frac{180°}{15} = 12° \text{ electrical degree}$$

the distribution factor for fundamental, $k_d = \dfrac{\sin\frac{nq\gamma}{2}}{q\sin\frac{n\gamma}{2}}$

$$K_{d1} = \frac{\sin\left(\frac{5\times 12°}{2}\right)}{5\sin\left(\frac{12°}{2}\right)} = 0.958$$

the distribution factor for third harmonic,

$$K_{d3} = \frac{\sin\left(\frac{3\times 5\times 12^{\circ}}{2}\right)}{5\sin\left(\frac{3\times 12^{\circ}}{2}\right)} = 0.647$$

the distribution factor for fifth harmonic,

$$K_{ds} = \frac{\sin\left(\frac{5\times 5\times 12^{\circ}}{2}\right)}{5\sin\left(\frac{5\times 12^{\circ}}{2}\right)} = 0.2$$

$$Total\ no.\ of\ turns\ perphase,\quad N_{ph} = \frac{90\times 4}{6} = 60 \text{ turns}$$

$$\text{the fundamental phase EMF, } E_{ph1} = 4.44\, K_{dl}\, K_{cl}\, f\, N_{ph}\, \Phi_1 \text{ V}$$

$$E_{phl} = 4.44(0.9569)\,(0.951)(50)(60)(0.015) = 181.82 \text{ V}$$

$$\text{the third component, } E_3 = E_1 \times \frac{\Phi_3}{\Phi_1} \times \frac{K_{d3}}{K_{d1}} \times \frac{K_{c3}}{K_{c1}} = 22.8 \text{ V}$$

$$\text{the fifth component, } E_5 = E_1 \times \frac{\Phi_5}{\Phi_1} \times \frac{K_{d5}}{K_{d1}} \times \frac{K_{c5}}{K_{c1}} = 0 \text{ V}$$

$$\text{The RMS phase voltage} = \sqrt{E_1 + E_3 + E_5} = \sqrt{(181.82)^2 + (22.8)^2 + 0}$$
$$= 183.24 \text{ V}$$

$$\text{The line voltage } E_L = \sqrt{3}\, E_{ph} = \sqrt{3} \times 183.24 = 317.38 \text{ V}$$

Example 4.6: A 10-pole, 600 rpm, 50 Hz, alternator has the following sinusoidal flux density distribution. $B = \sin\theta + 0.4\sin 3\theta + 0.2\sin 5\theta$ Wb/m^2. The alternator has 180 slots with two-layer three-turn coils with a coil span of 15 slots. The coils are connected in 60° groups. If the armature diameter is 1.2 m and core length is 0.4 m, calculate (a) the expression for instantaneous EMF/conductor (b) the expression for instantaneous EMF/coil (c) the phase and line voltages if the machine is star connected.

SOLUTION:

$$\text{Area under one pole pitch} = \pi\, DL/p = \pi \times 1.2 \times 0.4/10 = 0.1508 \text{ m}^2$$

$$\text{Fundamental flux/pole, } \Phi_l = \text{average flux density} \times \text{area}$$
$$= 2/\pi \times 1 \times 0.1508 = 0.096 \text{ Wb}$$

$$\text{(a) rms value of EMF induced/conductor} = 2.22 f\, \Phi_1 = 2.22 \times 50 \times 0.096 = 10.656 \text{ V}$$

$$\text{maximum value of EMF/conductor} = \sqrt{2} \times 10.656 = 15.07 \text{ V}$$

$$\text{3rd harmonic voltage} = 0.4 \times 15.07 = 6.02 \text{ Vs}$$

$$\text{5th harmonic voltage} = 0.2 \times 15.07 = 3.01 \text{ Vs}$$

the expression for instantaneous EMF/conductor $e = 15.07 \sin\theta + 6.02 \sin 3\theta + 3.01 \sin 5\theta$ Vs

(b) conductors/slot = 6 = conductors/coil, slots = 180, coil span = 15 slots

slots/pole = 18

slot angle = 180/number of slots/pole = 180/18 = 10°

coil is short chorded by 3 slots

hence $\varepsilon = 30^\circ$

Pitch factor $k_{pn} = \cos n\varepsilon/2$

$$k_{p1} = \cos\varepsilon/2 = \cos 30/2 = 0.9659$$

$$k_{p3} = \cos 3\times 30/2 = 0.707$$

$$k_{p5} = \cos 5\times 30/2 = 0.2588$$

Fundamental rms value of e.m.f. induced/coil = $4.44\, k_p\, f\, \Phi_1\, N_{ph}$

$$= 2.22\times 0.9659\times 50\times 0.096\times 6 = 61.76 \text{ V}$$

Maximum value of EMF induced/coil = $\sqrt{2}\times 61.76 = 87.34\ V$

Similarly, third harmonic voltage = 25.53 V

fifth harmonic voltage = 4.677 V

expression for instantaneous EMF/coile = 87.34 sinθ + 25.53 sin 3θ + 4.677 sin 5 θ V

slot angle γ = 180/number of slots/pole = 180/18 = 10°

number of slots/pole/phase = $180/(10\times 3) = 6$

Distribution factor $k_{dn} = (\sin q\, n\gamma/2)/(q \sin n\gamma/2)$

$$k_{d1} = \sin(6\times 10/2)/6\sin(10/2) = 0.956$$

$$k_{d3} = \sin(6\times 3\times 10/2)/6\sin(3\times 10/2) = 0.644$$

$$k_{d5} = \sin(6\times 5\times 10/2)/6\sin(5\times 10/2) = 0.197$$

$$N_{ph} = 180\times 6/6 = 180 \text{ Turns/phase}$$

value of EMF induced = $4.44\, k_{pn}\, k_{dn}\, (nf)\, \Phi_n\, N_{ph}$ for any nth harmonic

fundamental voltage $E_{ph1} = 4.44\, k_{p1}\, k_{d1}\, (f)\, \Phi_1\, N_{ph}$

$$= 4.44\times 0.9659\times 0.956\times 50\times 0.096\times 180 = 3542.68 \text{ V}$$

Similarly, third harmonic voltage $E_{ph3} = 697.65$ V

fifth harmonic voltage $E_{ph5} = 39.09$ V

$$\text{Phase voltage} = \sqrt{E_{ph1}^2 + E_{ph3}^2 + E_{ph5}^2}$$

$$= \sqrt{(3542.68)^2 + (697.652)^2 + (39.09)^2} = 3610.93 \text{ V}$$

$$\text{Line voltage} = \sqrt{3} \times \sqrt{E_{ph1}^2 + E_{ph5}^2} = \sqrt{3}\sqrt{(3542.68)^2 + (39.09)^2}$$

$$= 6136.48 \text{ V}$$

4.6 MAGNETIC FLUXES IN ALTERNATORS

There are three main fluxes associated with an alternator:

(a) Main useful flux linked with both field and armature winding.
(b) Leakage flux linked only with armature winding.
(c) Leakage flux linked only with field winding.

4.6.1 Armature Reaction

The useful flux, which links with both windings, is due to combined MMF of the armature winding and field winding. When the armature winding of an alternator carries current, an MMF sets in armature. This armature MMF reacts with field MMF producing the resultant flux, which differs from flux of field winding alone. The effect of armature reaction depends on nature of load (power factor of load). At no load condition, the armature has no reaction due to absence of armature flux. When armature delivers current at unity-power factor load, the resultant flux is displaced along the air gap toward the trailing pole tip. Under this condition, armature reaction has distorting effect on MMF wave as shown in Figure 4.11a at zero-lagging power factor loads the armature current is lagging by 90° with armature voltage. Under this condition, the position of armature conductor when inducing maximum EMF is the centerline of field MMF. Since there is no distortion but the two MMF are in opposition, the armature reaction is now purely demagnetizing as shown in Figure 4.11b. Now at zero leading power factor, the armature current leads armature voltage by 90°. Under this condition, the MMF of armature as well as the field winding are in same phase and additive.

The armature MMF has magnetizing effect due to leading armature current as shown in Figure 4.11c.

4.6.2 Armature Reaction Reactance (X_{ar})

In all the conditions of the load power factors, there is change in the terminal voltage due to the armature reaction. Mainly the practical loads are inductive in nature, due to demagnetizing effect of armature reaction, there is reduction in the terminal voltage. Now this drop in the voltage due to the interaction of armature and main flux. This drop is not across any physical element.

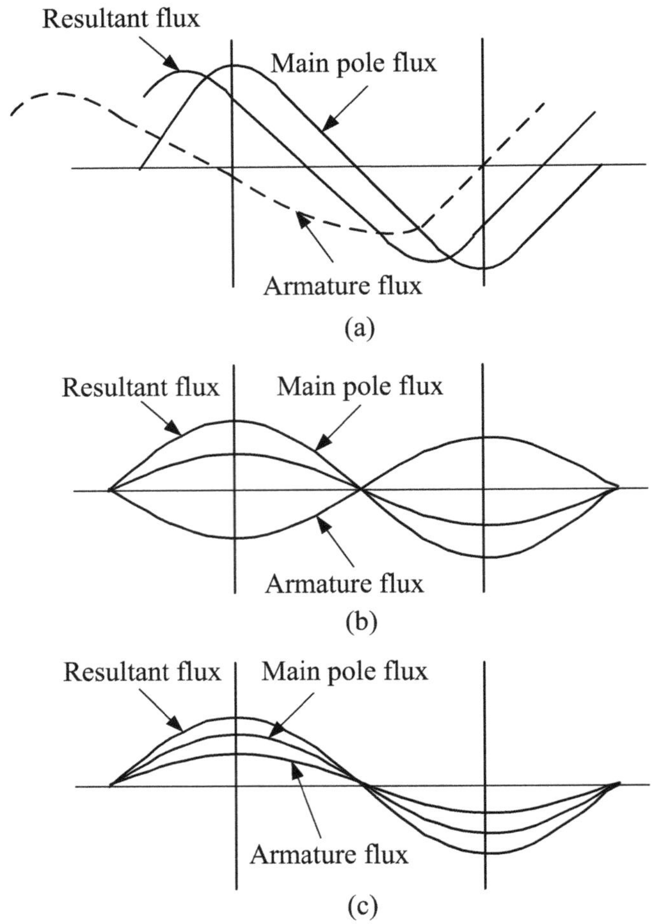

FIGURE 4.11 Demagnetizing and magnetizing effect of armature reaction, (a) Unity power factor, (b) Zero power factor leading, (c) Zero power factor lagging.

But to quantify the voltage drop due to the armature reaction, armature winding is assumed to have a fictitious reactance. This fictitious reactance of the armature is called armature reaction reactance denoted as Xar Ω/ph. And the drop due to armature reaction can be accounted as the voltage drop across this reactance as $I_{ar}X_{ar}$.

4.6.3 Armature Leakage Reactance

When armature carries a current, it produces its own flux. Some part of this flux completes its path through the air around the conductors itself. Such a flux is called leakage flux. This is shown in the Figure 4.12.

This leakage flux makes the armature winding inductive in nature. So winding possesses a leakage reactance, in addition to the resistance.

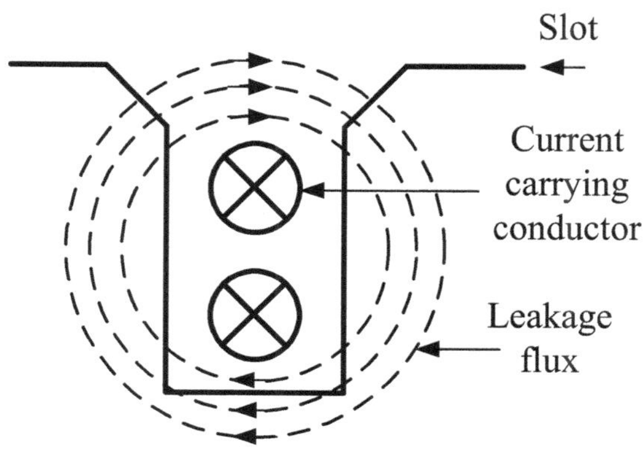

FIGURE 4.12 Armature leakage flux.

So if "L" is the leakage inductance of the armature winding per phase, then leakage reactance per phase is given by $X_{al} = 2\,\pi\ f\ L\ \Omega/\text{ph}$. The value of leakage reactance is much higher than the armature resistance. Similar to the DC machines, the value of armature resistance is very small.

4.7 CONCEPT OF SYNCHRONOUS REACTANCE AND IMPEDANCE

From the above discussion, it is clear that armature winding has one more parameter which is armature reaction reactance in addition to its resistance and the leakage reactance. The sum of the factious armature reaction reactance accounted for considering armature reaction effect and the leakage reactance of the armature is called synchronous reactance of the alternator denoted as X_s.

$$\text{so } X_s = X_{al} + X_{ar}\ \Omega/ph \tag{4.15}$$

As both X_{al} and X_{ar} are ohmic values per phase, synchronous reactance is also specified as ohms per phase.

Now from this, it is possible to define an impedance of the armature winding. Such an impedance obtained by combining per phase values of synchronous reactance and armature winding resistance is called synchronous impedance of the alternator denoted as Z_s.

$$\text{so} \qquad Z_s = r_a + j\ X_s \Omega/ph$$

$$\text{and} \qquad |Z_s| = \sqrt{r_a^2 + (X_s)^2}\ \Omega/ph$$

For getting a standard frequency, alternator is to be driven at synchronous speed. So word synchronous used in specifying the reactance and impedance is referred to the working speed of the alternator. Generally, impedance of the winding is constant but in case of the alternator, synchronous reactance depends on the load and its power factor condition, hence synchronous impedance also varies with the load and its power factor conditions.

4.8 EQUIVALENT CIRCUIT OF AN ALTERNATOR

Now it is clear that in all there are three important parameters of armature winding namely armature resistances r_a, leakage reactance X_{al}, and armature reaction reactance X_{ar}. If E_{ph} is induced EMF per phase on no load condition then on load it changes due to armature reaction as shown in Figure 4.13. As current I_a flows through the armature, there are two voltage drops across R_a and X_{al}, respectively. Hence finally terminal voltage v_t is less than E_r by amount equal to the drops across r_a and X_{al}.

In, practice, the leakage reactance X_{al} and the armature reaction reactance X_{ar} are combined to get synchronous reactance X_s. Hence the equivalent circuit of an alternator gets modified as shown in the Figure 4.14.

E_{ph} = Induced EMF per phase on no load
V_{tph} = Terminal voltage per phase on load
I_{aph} = Armature resistance per phase

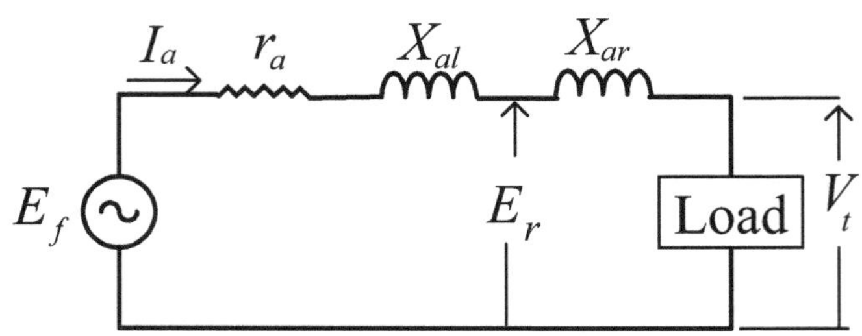

FIGURE 4.13 Equivalent circuit.

I_{aph} = Armature resistance per phase

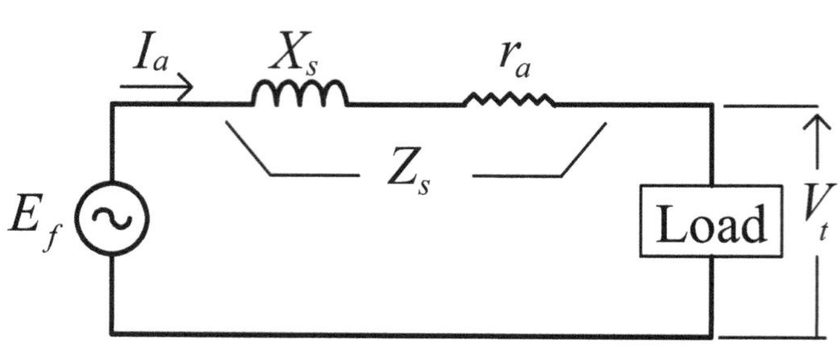

FIGURE 4.14 Equivalent circuit of an alternator.

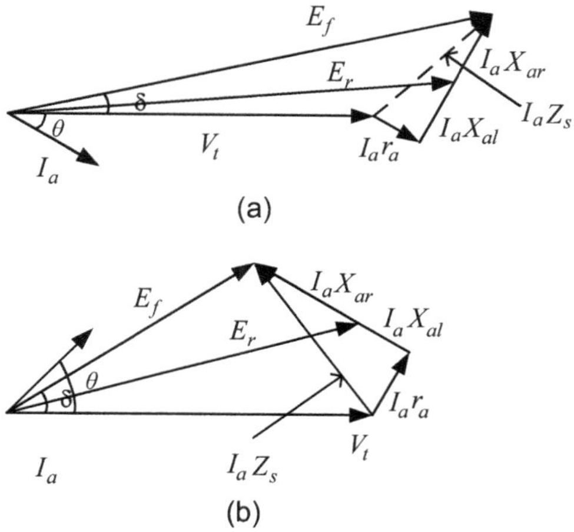

FIGURE 4.15 Phasor diagram of cylindrical rotor alternator at (a) lagging p.f. (b) leading p.f.

From the circuit diagram, it is clear that the open circuit voltage E_f is the summation of the terminal V_t and the drop across the synchronous impedance X_s and the armature resistance.

Hence,

$$E_f = V_t + I_a r_a + j\,I_a\left(X_{ar} + X_{al}\right) \tag{4.16}$$

where, $X_s = \left(X_{ar} + X_{al}\right)]$

$$E_f = V_t + I_a Z_s$$

where, $Z_s = \left\{r_a + j\,I_a\left(X_{ar} {}_{+} X_{al}\right)\right\}]$

Also, $E_f = E_r + j\,I_a\,X_{ar}$

where E_r is the EMF induced by the resultant flux Φ_r.

The Equation (3.16) can also be written as

$$E_f = \sqrt{\left(V\cos\theta + I_a r_a\right)^2 + \left(V\sin\theta \pm I_a X_s\right)^2}$$

where the positive sign for lagging power factor and the negative sign for leading power factor.

Figure 4.15a,b shows the phasor diagram of an alternator under lagging and leading p.f. operation respectively.

5 The Synchronous Generator Steady and State Performance

5.1 VOLTAGE REGULATION OF A GENERATOR

Voltage regulation of an alternator is defined as the change in the generator terminal voltage from its no-load value with the change of current I_a.

$$\text{Voltage regulation} = E_f - V_t \text{ [V]}$$

$$\begin{aligned} \%\text{ voltage regulation} &= \frac{\text{No loadvoltage-Full-load voltage}}{\text{Full-load voltage}} \times 100 \\ &= \frac{E_f - V_t}{V_t} \times 100 \end{aligned} \tag{5.1}$$

The term $E_f - V_t$ is the arithmetic difference and not the phasor difference. The factors affecting the voltage regulation of an alternator are:

(i) $I_a r_a$ drop in armature winding
(ii) $I_a X_{al}$ drop in armature winding
(iii) Voltage change due to armature reaction

The less value of regulation means a better generator.

We have seen that change in terminal voltage due to armature reaction depends upon the armature current as well as power-factor of the load. For leading load p.f., the no-load voltage is less than the full-load voltage. Hence, voltage regulation is negative in this case. The effects of different load power factors on the change in the terminal voltage with changes of load on the alternator are shown in Figure 5.1. Since the regulation of an alternator depends on the load and the load power factor, it is necessary to mention power factor while expressing regulation.

5.2 VECTOR DIAGRAM OF MACHINE MMF

The field MMF M_f is assumed to be sinusoidal by neglecting the field harmonics. M_f, I_f and φ_f are in phase. The armature reaction MMF. M_a is produced from the armature current I_a and in phase with I_a and φ_a, M_a is calculated as follows:

DOI: 10.1201/ 9781003795841-5

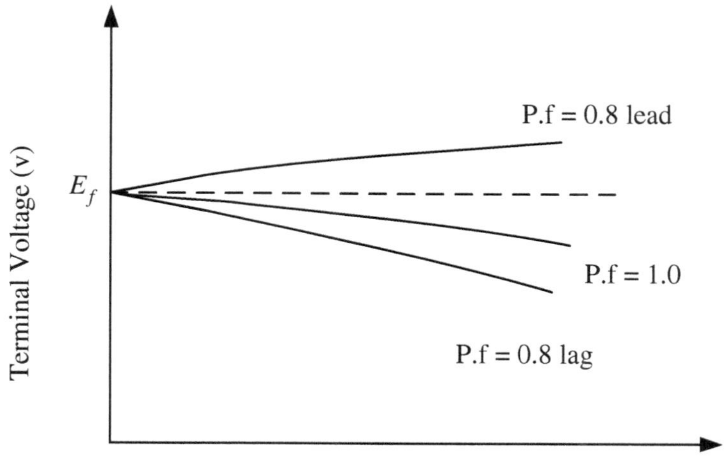

FIGURE 5.1 Voltage characteristics of cylindrical rotor alternator at constant field excitation.

$$M_a = \frac{2.71 I_a T_{ph}}{2P} \tag{5.2}$$

where T_{ph} is the no. of turns/phase.

The number 2.7 is due to the flux fundamental component.

The resultant field $\overline{M}_r$ is given as:

$$\overline{M}_r = \overline{M}_f + \overline{M}_a \tag{5.3}$$

$\overline{M}_r$ and $\overline{E}_r$ are perpendicular with each other and $\overline{E}_r$ lags $\overline{M}_r$ by 90°. The EMF $\overline{E}_f$ is induced in the armature in absence of φ_a and M_a. $\overline{E}_r$ is the resultant EMF and is true for unsaturated machine as shown in Figure 5.2b.

5.3 DETERMINATION OF VOLTAGE REGULATION

The ratings of commercial alternators are very high. It is neither convenient nor practicable to determine the voltage regulation by direct loading. There are several indirect methods of determining the voltage regulation of an alternator. These methods require only a small amount of power as compared to the power required for direct loading method. Four such methods are:

1. Synchronous impedance or EMF method
2. Ampere-turn or MMF method
3. AIEE method
4. Z.P.F. method

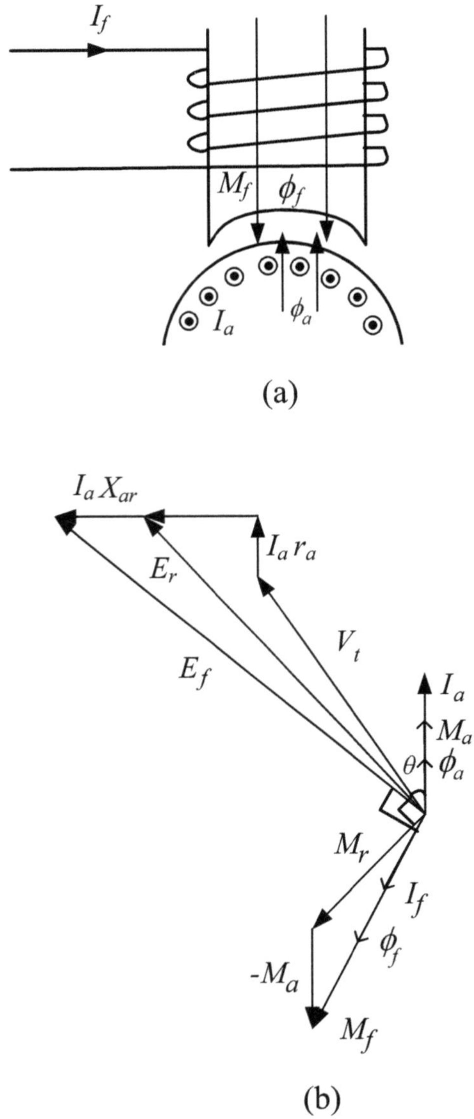

FIGURE 5.2 Phasor diagram for unsaturated synchronous generator.

For these method, the following data are required:

(i) Armature resistance
(ii) Open-circuit characteristic (*OCC*)
(iii) Short-circuit characteristic (*SCC*)

5.3.1 Synchronous Generator Tests

To obtain the parameters of a synchronous generator, we perform three simple tests as described below:

(i) The armature resistance r_a

Armature resistance per phase is determined by using direct current and the voltmeter-ammeter method. This is the DC value. The effective armature resistance (AC resistance) is greater than this value due to skin effect. It is a usual practice to take the effective resistance 1.5 times the DC value $\left(r_{\text{a}} = 1.2 \sim 1.8\, r_{\text{dc}}\right)$.

The reason behind that is the AC current tends to flow on the conductor surface. If we assume that the conductor is consists of large number of very thin conductors and each conductor has its flux lines around it. As a result of this flux, each conductor has it inductance. The inductance of the conductor at the center will be the highest and inductance decreases as the conductor is far from the center as shown in Figure 5.3. In this case, the inductance and of inner part of the conductor is higher than the inductance and reactance of the outer part, the current prefers to flow on the skin of the conductor. Hence, the effective cross-section area where the current flow is smaller than the conductor area which means that the resistance is higher.

(ii) Open-circuit characteristic (*OCC*)

Like the magnetization curve for a DC machine, the OCC of an alternator is the curve between armature terminal voltage (phase value) on open circuit and the field current when the alternator is running at rated speed. Figure 5.4a shows the circuit for determining the *OCC* of an alternator. The alternator is run on no-load at the rated speed. The field current if is gradually increased from zero (by adjusting field rheostat) until open-circuit voltage E_f (phase value) is about 50% greater than the rated phase voltage.

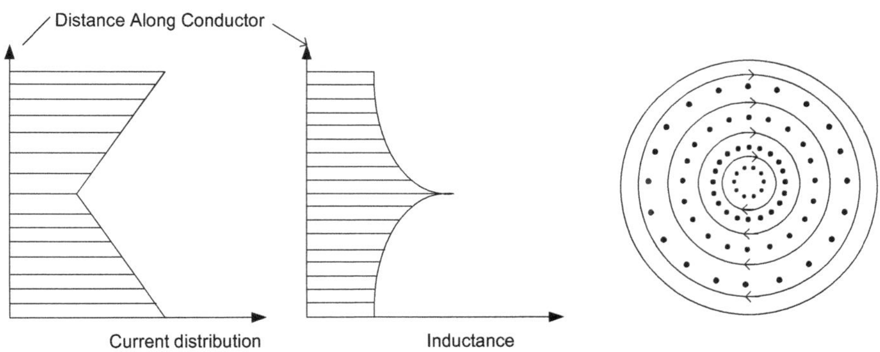

FIGURE 5.3 Skin effect.

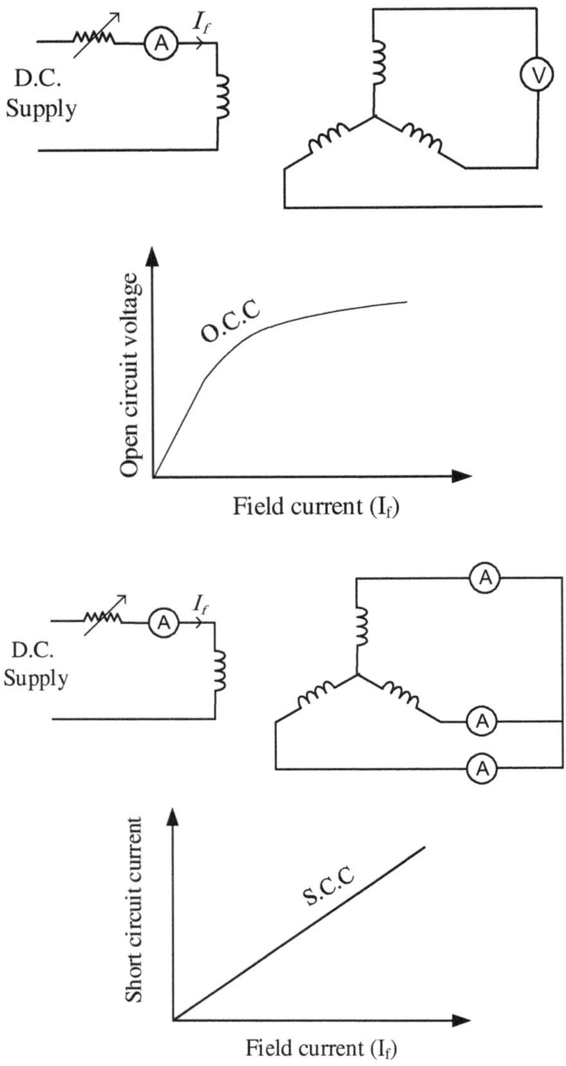

FIGURE 5.4 Circuit diagram to perform OCC and SC tests.

The graph is drawn between open-circuit voltage values and the corresponding values of I_f as shown in Figure 5.4a.

(iii) Short-circuit characteristic (*SCC*)

In a short-circuit test, the alternator is run at rated speed and the armature terminals are short-circuited through identical ammeters. Only one ammeter need be read; but three are used for balance. The field current If is gradually increased from zero until the short-circuit armature current I_{sc} is about twice the rated current. The graph between short-circuit armature current and field current gives the *SCC* as shown in Figure 5.4b. There is no need to take more than one reading because *SCC* is a straight line passing through the origin. The reason is simple. Since armature resistance is much smaller than the synchronous reactance, the short-circuit armature current lags the induced voltage by very nearly 90°. Consequently, the armature flux and field flux are in direct opposition and the resultant flux is small. Since the resultant flux is small, the saturation effects will be negligible and the short-circuit armature current, therefore, is directly proportional to the field current over the range from zero to well above the rated armature current.

At any value of field current, if E_f is the open circuit voltage and I_{sc} is the short circuit current, then for this value of excitation:

$$Z_s = E_f(\text{open circuit})/I_a\left(\text{short circuit}\right) \tag{5.4}$$

Because of the nonlinear nature of the OCC, the ratio of open-circuit voltage and short-circuit current at various values of excitation currents are different. If the value of Z_s at different excitation are calculated and plotted, the Z_s curve is obtained as shown in Figure 5.5. It is obvious that when the *OCC* is linear (unsaturated region), the value of Z_s is constant. At higher values of field current, saturation increases and the synchronous impedance decreases. The value of Z_s calculated for the unsaturated region of the *OCC* is called the unsaturated value of the synchronous impedance.

If r_a is the effective resistance of the armature per phase, the synchronous reactance X_s is given by:

$$X_s = \sqrt{(Z_s)^2 - (r_a)^2} \tag{5.5}$$

If V_t is the magnitude of the rated voltage of the machine and the regulation is to be calculated for a load current I at a power factor angle Φ, then the corresponding magnitude of the open circuit voltage E_0 is

$$E_f = V_t + I_a Z_s \tag{5.6}$$

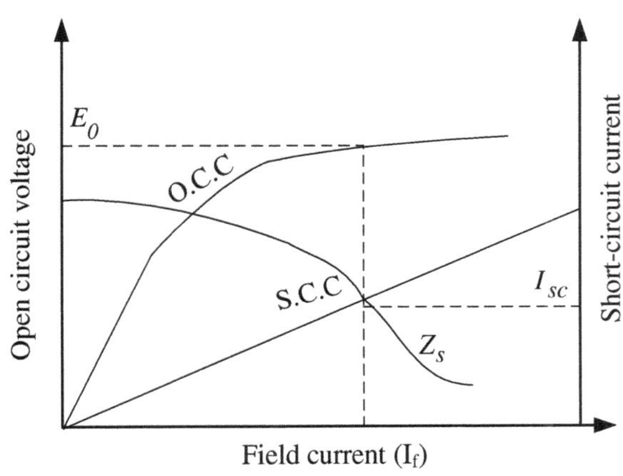

FIGURE 5.5 Synchronous impedance variation with the armature current.

From the OCC and SCC, The armature reaction M_a component can be obtained in the laboratory instead of using equation (5.2). By drawing E_r on the vertical axis of the OCC, the M_r component is then obtained. By locating M_f on the horizontal axis, the open circuit EMF E_f is obtained as shown in Figure 5.6b. The component $I_a X_{la}$ is locating in the vertical axis and used to complete Potier triangle, which is proportional to the armature current I_a since the vertical line is $I_a X_{la}$ and the base is M_a. In short circuit, the machine appears to be loaded by a pure inductance X_{la} as r_a is usually very small compared with X_{la}. The vector diagram in case of short circuit is shown in Figure 5.6a, the field MMF M_f in this case may be large enough to overcome the armature reaction MMF M_a and to produce a voltage which can cancel $I_a X_{la}$.

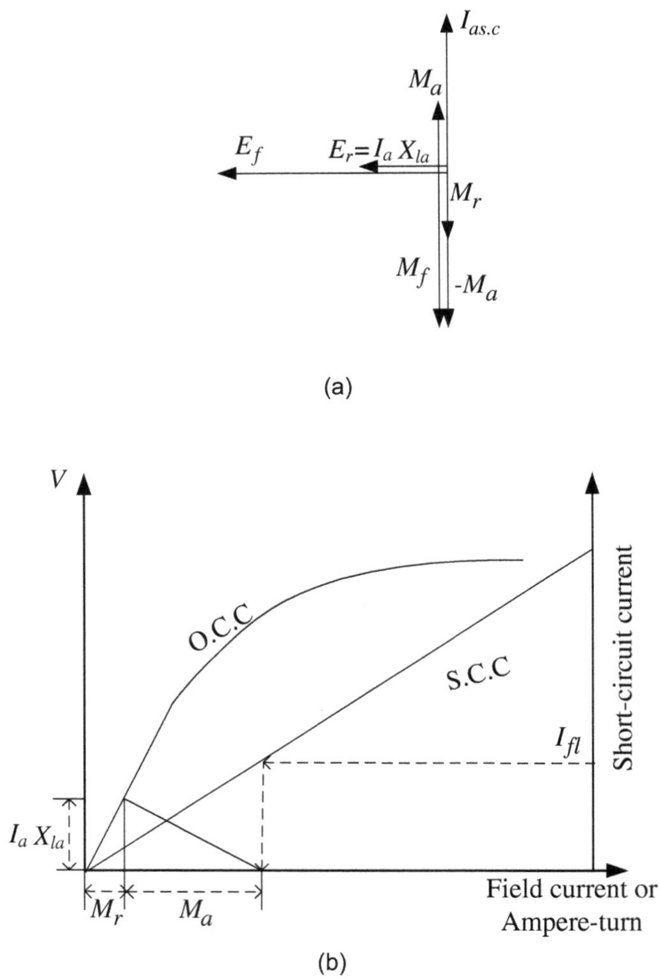

FIGURE 5.6 Vector diagram of short-circuited synchronous generator.

5.3.2 Synchronous Impedance or EMF Method

The synchronous impedance triangle must first be obtained from the OCC and SCC, as already explained, after which the regulation is derived by the aid of the vector diagrams shown in Figure 5.4. Let OV be the terminal voltage (or phase voltage in the case of a 3-phase alternator) and let OI_a represent full-load current. The voltage-drop triangle, obtained from the armature resistance and synchronous reactance, is then erected in position, the synchronous reactance voltage-drop vector being at right angles to the current vector. The induced voltage is then given by OE, and the % regulation by $\frac{OE-OV}{OV}\times 100$ percent. The construction can be carried out for any phase angle, and also for any value, of the current.

Figures 5.7a,b illustrates cases of lagging and leading armature currents respectively. It will be observed that a lagging current causes a greater voltage to rise on throwing off the load than does a current at unity power factor, while a leading current brings about a reduced voltage rise. In fact, if the current be made to lead by a sufficient angle, the voltage on no-load may actually be lower than the voltage on load.

$$E_f=\sqrt{\left(V_t\cos\theta+I_ar_a\right)^2+\left(V_t\sin\theta-I_aX_s\right)^2} \tag{5.7}$$

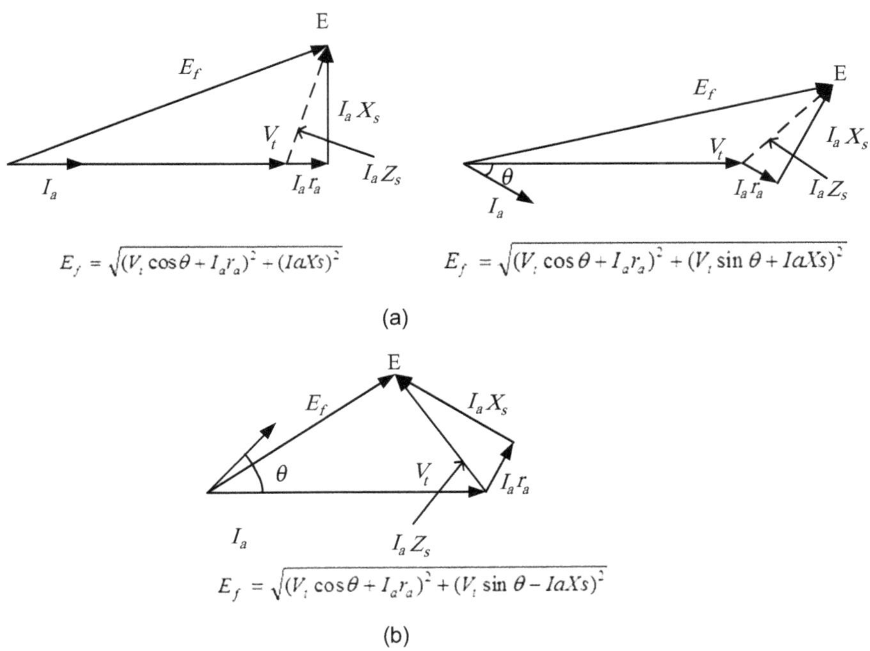

FIGURE 5.7 Voltage regulation for different power factor.

Example 5.1: A 3-phase star connected alternator is rated at 100 kVA. On a short-circuit a field current of 50 amp gives the full load current. The EMF generated on open circuit with the same field current is 1,575 V/phase. Calculate the voltage regulation at (a) 0.8 power factor lagging, and (b) 0.8 power factor Lagging. Assume armature resistance is 1.5 Ω.

SOLUTION:

Let the rated terminal voltage of the alternator

$$V = 1575 \text{ V/phase}$$

$$\therefore \therefore \text{Full load current} \qquad I = \frac{1000\times10^3}{3\times1575} = 211.64 \text{ A}$$

Synchronous impedance

$$Z_s = \frac{O.C\ voltage}{S.C.\ current} \text{ for same field excitation}$$

$$\text{Or} \qquad Z_s = \frac{1575}{211.64} = 7.442\ \Omega$$

$$X_s = \sqrt{\left(Z_s^2 - R_a^2\right)} = \sqrt{(7.442)^2 + (1.5)^2} = 7.289\ \Omega$$

(a) At lagging power factor, the no load voltage E_f is given by the equation

$$E_f = \sqrt{\left(V_t\cos\theta + I_aR_a\right)^2 + \left(V_t\sin\theta + I_aX_s\right)^2}$$
$$= \sqrt{(1575\times0.8+211.64\times1.5)^2 + (1575\times0.6+211.64\times7.289)^2}$$
$$= 2945.63 \text{ V}$$

$$\therefore \%\ \text{Regulation} = \frac{2945.63-1575}{1575}\times100 = 87.02\%$$

(b) At leading power factor, the no load voltage E_f is given by the equation

$$E_f = \sqrt{\left(V_t\cos\theta + I_aR_a\right)^2 + \left(V_t\sin\theta - I_aX_s\right)^2}$$
$$= \sqrt{(1575\times0.8+211.64\times1.5)^2 + (1575\times0.6-211.64\times7.289)^2}$$
$$= 1686.878 \text{ V}$$

$$\therefore \%\ \text{Regulation} = \frac{1686.878-1575}{1575}\times100 = 7.1\%$$

Example 5.2: 20 MVA, 13 kV, 3-phase, Y-connected alternator has an armature resistance of 0.2 Ω/phase and a synchronous armature reactance of 1.2 Ω/phase. Find the full load generated voltage per phase at:

(a) Unity power factor
(b) a power factor of 0.7 lagging
(c) a power factor of 0.95 leading

Then calculate then voltage regulation for the 3 power factor conditions.

Solution:

$$V_t = \frac{13000}{\sqrt{3}} = 7505.8 \text{ V}$$

$$I_a = \frac{MVA \times 1000000}{3V} = \frac{20 \times 1000000}{3 \times 7505.8} = 888.25 \text{ A}$$

$$I_a R_a = 888.25 \times 0.2 = 177.65 \text{ V}$$

$$I_a X_s = 888.25 \times 1.2 = 1065.9 \text{ V}$$

(a) at unity power factor

$$E_f = V_t + I_a Z_s = (7505.8 + 177.65) + j1065.9$$
$$= 7757\angle 7.9° \text{ V}$$

$$\therefore \%\text{Regulation} = \frac{7757 - 7505.8}{7505.8} \times 100 = 3.35\%$$

(b) at 0.7 lagging power factor

$$E_f = \sqrt{(V_t \cos\theta + I_a R_a)^2 + (V_t \sin\theta + I_a X_s)^2}$$
$$= \sqrt{(7505.8 \times 0.7 + 177.65)^2 + (7505.8 \times 0.714 + 1065.9)^2}$$
$$= 8413.36 \text{ V}$$

$$\therefore \% \text{ Regulation} = \frac{8413.36 - 7505.8}{7505.8} \times 100 = 12\%$$

(c) at 0.95 leading power factor

$$E_f = \sqrt{(V_t \cos\theta + I_a R_a)^2 + (V_t \sin\theta - I_a X_s)^2}$$
$$= \sqrt{(7505.8 \times 0.95 + 177.65)^2 + (7505.8 \times 0.312 - 1065.9)^2}$$
$$= 7419 \text{ V}$$

$$\therefore \% \text{ Regulation} = \frac{7419 - 7505.8}{7505.8} \times 100 = -1.15\%$$

The preceding example indicates the following interesting remarks:

1. The rising of power factor from lagging to unity power factor is still insufficient to produce 0% voltage regulation (because of the voltage drop in the resistive component).
2. At leading power factor loads, the armature reaction is magnetizing, and tends to produce additional generated voltage as the load is increased (since the no-load voltage E_f is less than the full load voltage V_t)
3. At lagging power factor, the armature reaction is demagnetizing, and results in a rapid decrease in generated voltage as the load is increased. However, in practice, the alternator terminal voltage is maintained constant by means of external voltage regulators which automatically increase or decrease the dc field excitation.
4. The difference between the generated voltage, E_f, and the terminal voltage, E_f, of an alternator is the synchronous impedance voltage drop $I_a Z_s$. The synchronous impedance and effective resistance per phase are determined by means of specific tests in a technique called the synchronous impedance method.

Example 5.3: A 1,000 kVA, 11 kV, 3-phase, Y-connected alternator is tested in accordance with the synchronous impedance technique. The data obtained is as follows:

(a) DC resistance test: V between lines = 12 DC V,
line current = 20 A

(b) Open-circuit test: Field current = 20 A
V between lines = 420 ac V

(c) Short-circuit test: Field current = 20 A
Line current = rated value

Calculate:
(1) Effective resistance, synchronous impedance and reactance per phase.
(2) Voltage regulation of the alternator at 0.85 lagging and 0.85 leading power factor.

SOLUTION:

$$(1)\ I_a = \frac{kVA \times 1000}{\sqrt{3}V} = \frac{1000 \times 1000}{\sqrt{3} \times 11 \times 1000} = 52.5\ \text{A}$$

$$\text{From dc test } r_{dc} = \frac{1}{2}\frac{12}{20} = 0.3\ \Omega$$

The ac resistance

$$\begin{aligned} r_a &= r_{dc} \times \text{skin effect factor} \\ &= 0.3 \times 1.5 = 0.45\ \Omega \end{aligned}$$

From open circuit and short circuit tests:

$$Z_s = \frac{opencircuit\ phase\ voltage}{short\ circuit\ current} \quad \text{(measured at the same field current)}$$

$$Z_s = \frac{\left(420/\sqrt{3}\right)}{52.5} = 4.619\Omega$$

$$X_s = \sqrt{\left(Z_s^2 - r_a^2\right)} = \sqrt{(4.619)^2 + (0.45)^2} = 4.61\ \Omega$$

$$(2)\ V_p = \frac{11000}{\sqrt{3}} = 6351\ \text{V}$$

$$(3)\ I_a r_a = 52.5 \times 0.45 = 23.6\ \text{V}$$

$$(4)\ I_a X_s = 52.5 \times 4.61 = 242\ \text{V}$$

at 0.85 lagging power factor

$$\begin{aligned} E_f &= \sqrt{\left(V_t \cos\theta + I_a r_a\right)^2 + \left(V_t \sin\theta + I_a X_s\right)^2} \\ &= \sqrt{\left(6351 \times 0.85 + 23.6\right)^2 + \left(6351 \times 0.527 + 242\right)^2} \\ &= 6501\ \text{V} \end{aligned}$$

$$\therefore \%\ \text{Regulation} = \frac{6501 - 6351}{6351} \times 100 = 2.37\%$$

(d) at 0.85 leading power factor

$$\begin{aligned} E_f &= \sqrt{\left(V_t \cos\theta + I_a r_a\right)^2 + \left(V_t \sin\theta - I_a X_s\right)^2} \\ &= \sqrt{\left(6351 \times 0.85 + 23.6\right)^2 + \left(6351 \times 0.527 - 242\right)^2} \\ &= 6248\ \text{V} \end{aligned}$$

$$\therefore \%\ \text{Regulation} = \frac{6248 - 6351}{6351} \times 100 = -1.62\%$$

5.3.3 The Magneto Motive Force MMF (Ampere-Turn) Method

In this method, the magneto motive force is used to calculate the voltage regulation rather than the generated voltage since the EMF method is assumed that the machine is unsaturated. The OCC and SCC are also used to obtain the field currents which produce both the rated voltage and current. From open circuit and short circuit characteristic, the armature reaction MMF M_a is obtained. After drawing the vector diagram, the resultant MMF M_r and M_a are used to determine the total ampere-turn

M_f. From the OCC, the open circuit voltage E_f is obtained by drawing a vertical line from the value M_f as shown in Figure 5.8a,b.

Then percentage voltage regulation can be determined by

$$\% VR = \frac{E_f - V_t}{V_t} \times 100$$

5.3.4 AIEE Method

The calculation of voltage regulation using the EMF method is unrealistic because the assumption of unsaturated magnetic circuit. Also, the MMF method are giving

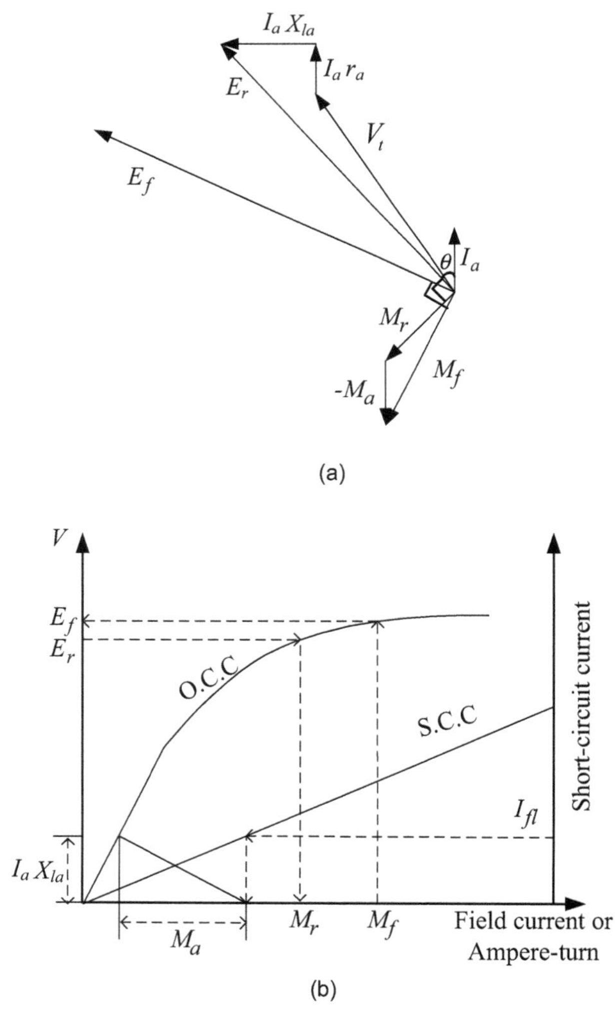

FIGURE 5.8 The MMF voltage regulation.

unrealistic value of regulation due to the same assumption. Because of their simplicity, these methods are being used in spite of their short coming. Therefore, AIEE has modified MMF method for calculation of regulation. The first step is to draw the vector diagram on the OCC and get the air gap and saturation currents (I_{fs} and I_{fg}) as shown in Figure 5.9. Using the obtained currents, the MMF diagram is drawn to get the total field current I_{ft}. This current is used again in the OCC to get the open circuit voltage E_f as explained in Figures 5.9 and 5.10.

5.3.5 Zero Power Factor Characteristics

Zero power factor is a relation between alternator terminal load voltage $\left(V_f\right)$ and field current $\left(I_f\right)$ when the machine is driven at its synchronous speed N_s, with its terminals connected to a pure inductive load. The armature current I_a must be kept constant during the test by changing the load. To plot ZPF characteristics only two points are required. One point is corresponding to the zero voltage and rated current (point P) that can be obtained from S.C.C. and the other is obtained by drawing $I_a X_{al}$ vertically. This zero power factor curve appears like *OCC* but shifted by a factor $I_a X_{al}$ vertically and horizontally by armature reaction MMF as shown in Figure 5.11. The ZPF curve can be completed by moving Potier triangle parallel to the triangle OQP so as its head on the OCC and its sides have a constant dimensions.

5.3.6 Zero Power Factor Method

During the operation of the alternator, resistance voltage drop $I_a r_a$ and armature leakage reactance drop $I_a X_{al}$ are actually EMF quantities and the armature reaction

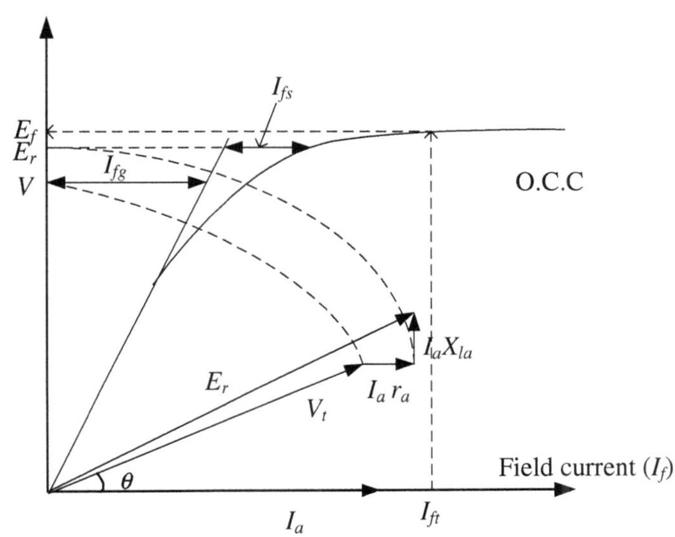

FIGURE 5.9 The AIEE vector diagram

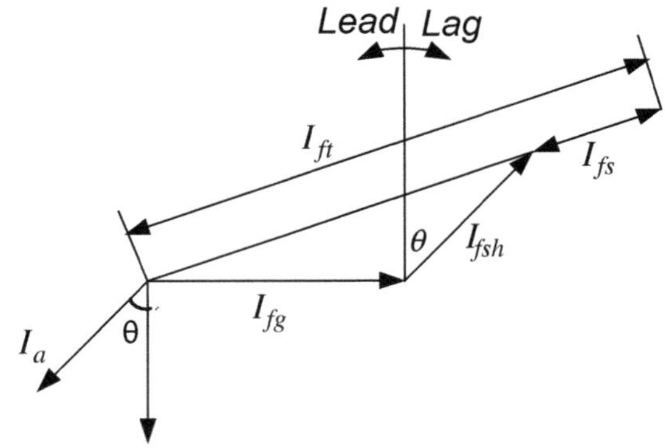

FIGURE 5.10 The AIEE phasor diagram.

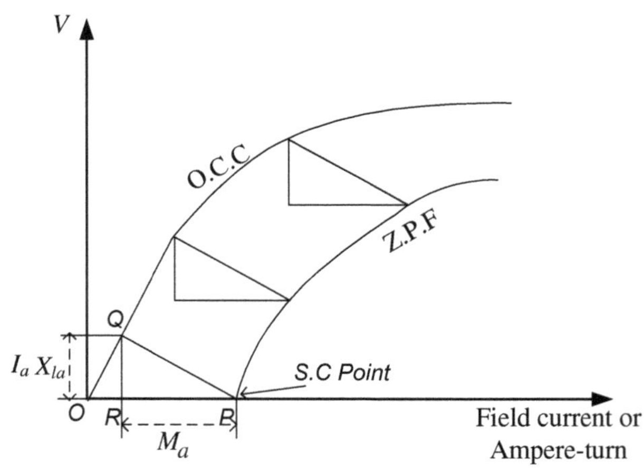

FIGURE 5.11 OCC, ZPF, and Potier triangle.

reactance is an MMF quantity. To determine the regulation of the alternator by the ZPF method, *OCC*, *SCC*, and *ZPF* test details and characteristics are required. Using these characteristics, draw a horizontal line at the rated voltage (V_t) meet the ZPF at P^* as shown in Figure 5.12. Take the length *OP* on the rated voltage line to get $O \times P^*$. From O^* draw a line parallel to the air gap line and meet the OCC at Q^*. By drawing the vector diagram, the no load E_f is obtained, and the voltage regulation can be calculated using the phasor diagram as shown in Figure 5.13.

The voltage regulation can be obtained by:

$$\% \, VR = \frac{E_f - V}{V} \times 100$$

Example 5.4: The data obtained on a 100 kVA, 1,100 V, 3-phase alternator is: DC resistance test, E between line = 6 V dc, I in lines = 10 A dc. Open circuit test, field

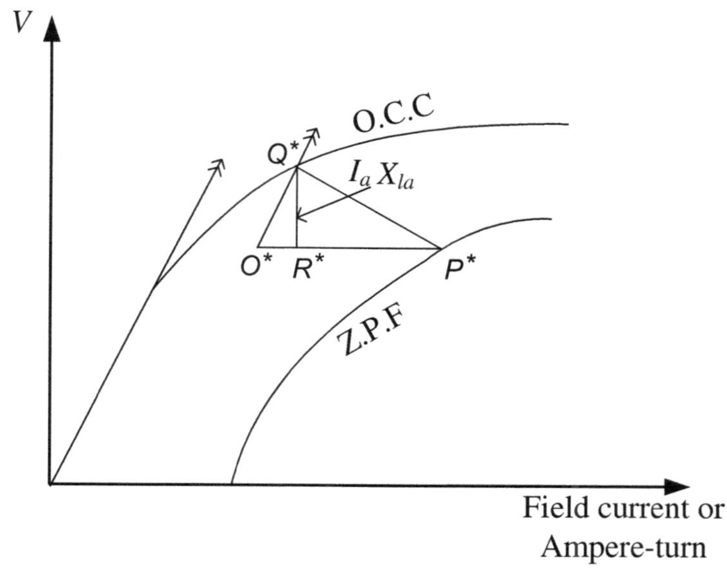

FIGURE 5.12 Constructing Potier triangle.

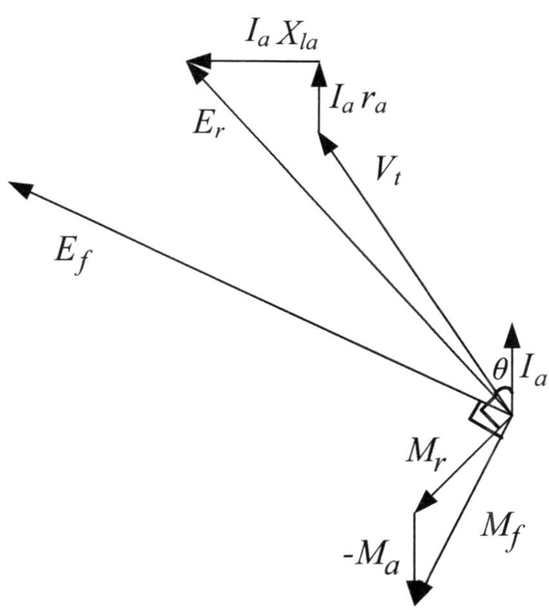

FIGURE 5.13 Phasor diagram.

current = 12.5 A dc, line voltage = 420 V ac. Short circuit test, field current = 12.5 A, line current = rated value, calculate the voltage regulation of alternator at 0.8 p.f. lagging.

Solution

Assume alternator to be star-connected, as usually

$$V_p = \frac{1100}{\sqrt{3}} = 635.1 \text{ V}$$

$$I_a = \frac{kVA \times 1000}{\sqrt{3}V} = \frac{100 \times 1000}{\sqrt{3} \times 1100} = 52.5 \text{ A}$$

$$r_{dc} = \frac{1}{2}\frac{12}{20} = 0.3\ \Omega$$

The ac resistance

$$r_a = r_{dc} \times \text{skin effect factor}$$
$$= 0.3 \times 1.5 = 0.45\ \Omega$$

From open circuit and short circuit tests:

$$Z_s = \frac{opencircuit\ phasevoltage}{short\ circuit\ current} \quad \text{(measured at the same field current)}$$

$$Z_s = \frac{(420/\sqrt{3})}{52.5} = 4.62\ \Omega$$

$$X_s = \sqrt{(Z_s^2 - r_a^2)} = \sqrt{(4.62)^2 + (0.45)^2} = 4.61\,\Omega$$

at 0.8 lagging power factor

$$E_f = \sqrt{(V_t \cos\theta + I_a r_a)^2 + (V_t \sin\theta + I_a X_s)^2}$$
$$= \sqrt{(635.1 \times 0.8 + 52.5 \times 0.45)^2 + (635.1 \times 0.6 + 52.5 * 4.61)^2}$$
$$= 820 \text{ V}$$

$$\therefore \%\ \text{Regulation} = \frac{820 - 635.1}{635.1} \times 100 = 29.11\%$$

Example 5.5: A 3-phase, 1,500 kVA, star-connected, 50 Hz, 200 V alternator has a resistance between each pair of terminals as measured by direct current is 0.16 Ω. Assume that the effective resistance is 1.5 times the Ohmic resistance. A field current of 70 A produces a short-circuit current equal to full-load-current of 376 A in each line. The same field current produces an EMF of 700 V on open circuit. Determine the synchronous reactance of the machine and its full load regulation at 0.8 power factor lagging.

SOLUTION:

$$Z_s = \frac{opencircuit\ phasevoltage}{short\ circuit\ current} \quad \text{(measured at the same field current)}$$

$$Z_s = \frac{\left(700/\sqrt{3}\right)}{376} = 1.075\,\Omega$$

$$\text{Ohmic resistance per phase} = \frac{0.16}{2} = 0.08\,\Omega$$

The effective resistance

$$r_a = 0.08 \times 1.5 = 0.12\ \Omega$$

Synchronous reactance

$$X_s = \sqrt{\left(Z_s^2 - r_a^2\right)} = \sqrt{(1.075)^2 + (0.12)^2} = 1.068\ \Omega$$

$$S_{3\text{-phase}} = \sqrt{3}V_L I_L$$

$$\therefore I_L = \frac{1500 \times 1000}{\sqrt{3} \times 2300} = 376\text{ A}$$

$$V_p = \frac{2300}{\sqrt{3}} = 1327.9\text{ V}$$

at 0.8 lagging power factor

$$E_f = \sqrt{\left(V_t \cos\theta + I_a r_a\right)^2 + \left(V_t \sin\theta + I_a X_s\right)^2}$$

$$= \sqrt{\left(1327.9 \times 0.8 + 376 * 0.12\right)^2 + \left(1327.9 \times 0.6 + 376 * 1.068\right)^2}$$

$$= 1631.7\text{ V}$$

$$\therefore \%\ \text{Regulation} = \frac{1631.7 - 1327.9}{1327.9} * 100 = 22\%$$

5.4 POWER-ANGLE CHARACTERISTICS

The performance of a synchronous machine can be represented by expressions of electrical output power (alternator case) and electromagnetic torque (motor case) as a function of the power (or torque) angle δ.

Assuming the alternator case, the generated current is expressed as:

$$I_a = \frac{E_f - V_t}{Z_s} \tag{5.8}$$

By assuming V_t V_t as a reference phasor, Eqn. (5.8) will be:

$$I_a = \frac{E_f}{Z_s}\angle\delta - \alpha - \frac{V_t}{Z_s}\angle -\alpha \tag{5.9}$$

Where $\alpha = \arctan\frac{x_s}{r_a}$

The power delivered by the alternator is given by:

$$P = 3\,V_t I_a \cos\theta \tag{5.10}$$

But since V_t is used as a reference for Eqn. (5.7) $I_a \cos\theta$ is the real component of Eqn. (5.10); thus

$$I_a \cos\theta = \frac{E_f}{|Z_s|}\cos(\delta - \alpha) - \frac{V_t}{|Z_s|}\cos(-\alpha) \tag{5.11}$$

Substituting Eqn. (5.11) into Eqn. (5.10) will give:

$$P = \frac{3E_f V_t}{|Z_s|}\cos(\delta - \alpha) - \frac{3V_t^2 r_a}{|Z_s|^2} \tag{5.12}$$

Figure 5.14 shows equation (5.12) plotted with power angle δ as an independent variable. From Figure 5.13, it is noticed that the maximum electrical power output power of an alternator occurs at angle less than 90 electrical degrees due to the effect of armature resistance r_a. Now, if r_a has a very small value so that it can be neglected, then the alternator eqn. (5.12) will be:

$$P = \frac{3E_f V_t}{|Z_s|}\sin\delta \tag{5.13}$$

A plot of equation (5.13) is shown in Figure 5.15, assuming $|E_f|$ and $|V_t|$ are constant. This figure illustrates two important points:

1. There is a definite upper limit to the power capability of a synchronous machine.
2. When δ becomes negative, power flow reverses. Power then flows into the electrical terminals, indicating that the machine is acting as a motor. In this case, developed torque of the motor is given by:

$$T_d = \frac{P}{\omega_s} = \frac{3}{\omega_s}\frac{|E||V_t|}{x_s}\sin\delta \tag{5.14}$$

where, ω_s is the synchronous speed.

Example 5.6: A 3-phase, 80 kW, star-connected, 400 V synchronous alternator has asynchronous impedance per phase of $0.1 + j1\,\Omega$. If the dc excitation is equivalent to 440 line Vs, calculate the line current, power factor, and efficiency when operating at full load, assuming no load losses of 2 kW.

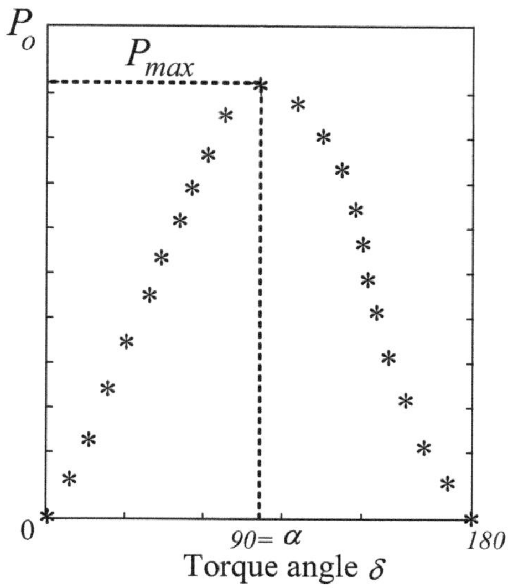

FIGURE 5.14 Power-load angle characteristics for an alternator.

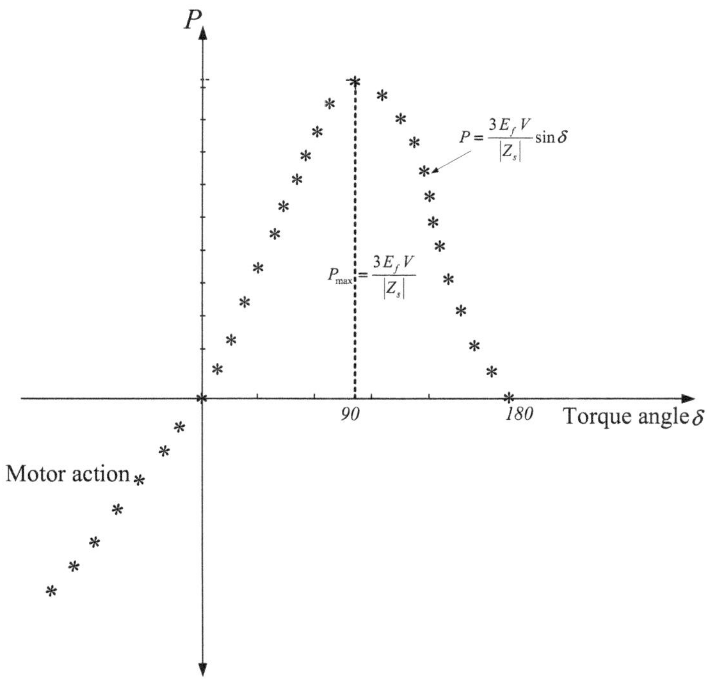

FIGURE 5.15 Power-load angle characteristics for E_f and V_t constant.

SOLUTION:

$$\alpha = \sin^{-1}\frac{r_a}{Z_s} = \sin^{-1}\frac{0.1}{\sqrt{0.1^2+1^2}} = \sin^{-1}\frac{0.1}{1.005} = 5.7°$$

Using equation (5.11)

$$80000 = \frac{(400/\sqrt{3})(400/\sqrt{3})^2}{1.005}\sin(\delta-5.7) - \frac{(400/\sqrt{3})}{1.005}\cos(-5.7)$$

$$0.457 = \sin(\delta-5.7) - 0.0904$$

$$\delta = 33.2+5.7 = 38.9°$$

Taking V as a reference phasor

$$E_{gp} = \frac{440}{\sqrt{3}}(\cos\delta + j\sin\delta) = \frac{440}{\sqrt{3}}(\cos 38.9 + j\sin 38.9)$$

$$E_{gp} = 254(0.778 + j0.628) = 197.6 + j159.5$$

$$\therefore I_a = \frac{E_{gp}-V_t}{V_t} = \frac{197.6 + j159.5 - 230.9}{0.1+j1} = 162.13\angle -6.099 \text{ A}$$

$$\text{The power factor} = \cos 6.099 = 0.994 \text{ lagging}$$

The efficiency

$$= \frac{80\times 1000}{80\times 1000 + 2000 + 3I_a^2 r_a} = \frac{80000}{82000 + 3\times 162.13^2 \times 0.1} = 89\%$$

5.5 PERFORMANCE CIRCLE DIAGRAM

When an alternator is connected to an " infinite busbar" (if a system is so large that it can absorb all the power an alternator can generate or supply all the power a machine would ever require as a motor, without any change in system frequency or voltage at the machine terminals, we can state that the system is an infinite busbar system), two variables may be used to control its performance; these are the excitation E_f and the machine input power.

First, consider the case when the excitation E_f is constant while the output (or input) power is varying. Consider the armature equation (5.8), the locus of current can be obtained for variable load power. The phasor E_f, although constant in magnitude, varies in phase with respect to V_t in accordance with the necessary changes in load angle. Since a constant E_f will trace out a circular locus, the same is true for $\frac{E_f}{Z_s}$.

And since $\frac{V_t}{Z_s}$ is fixed, the current phasor has the same circular locus as $\frac{E_f}{Z_s}$.

Although the case where E_f is numerically greater than V_t will be considered, the same applies for reduced excitations. For this condition equation (5.13) is presented by Figure 5.16.

The operating range of a synchronous alternator is the vector bounded by the reference line (RL) which coincides with the phasor V_t, and the unity power factor (UPF) line. In this range the power angle δ varies between zero (which represents the no-load condition since I_a is equal to zero) and (90 – α) electrical degree which represents the upper limit to the power capability of an alternator. For the sector above RL, the power angle δ is always positive, which means generator case. And for the sector below RL, the power angle δ is always negative, which means that the synchronous delivers negative power, and the machine is receiving power, that is, the machine is operating as a motor. The unit used for drawing any phasor in Figure 4.15 is the V. Now, in order to transfer these average units into power units, each phasor is multiplied by $\frac{|3V_t|}{|Z_s|}$. It must be realized that for a given synchronous machine, both V_t and Z_s are fixed quantities, which means that each phasor in Figure 5.16 is multiplied by a scalar quantity for converting the V unit into Watt.

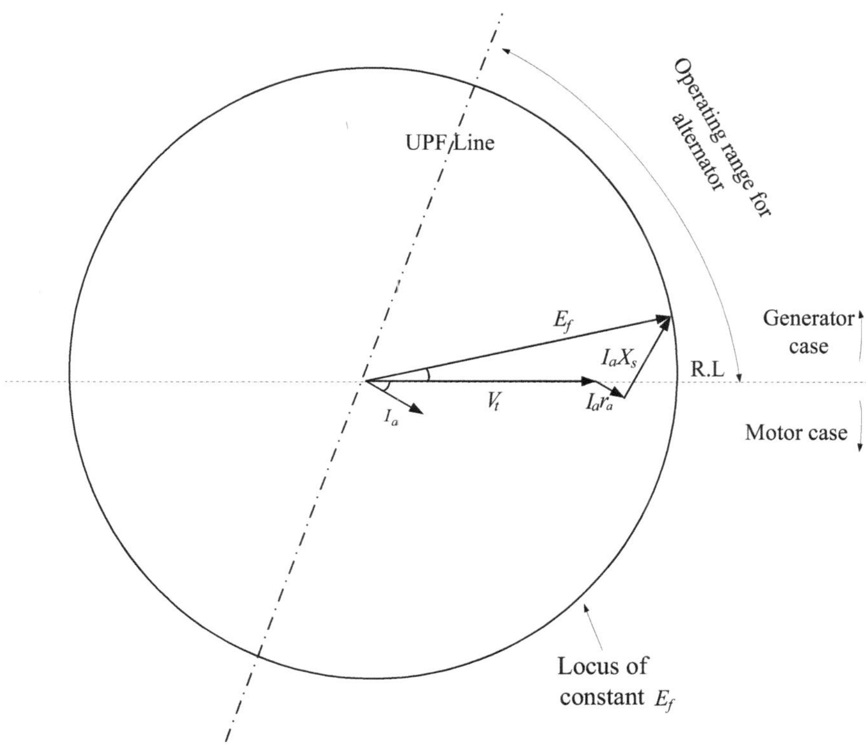

FIGURE 5.16 The locus diagram for constant E_f conditions.

Second, consider the case when the input power is constant while the excitation E_f is varying.

The input power to the alternator can be obtained from the following expression:

$$P_i = P_o + 3I_a^2 r_a \tag{5.15}$$

If r_a is so small that it can be neglected, equation (5.15) will be

$$P_i = P_o = 3V_t I_a \cos\theta = 3\frac{V}{Z_s} I_a Z_s \cos\theta \tag{5.16}$$

$$= \text{Scalar} \times \text{projection of } I_a Z_s \text{ phasor on the UPF line.}$$

The input power $P_i = (\text{length ab}) \times \text{scale of drawing} \times 3\frac{V_t}{Z_s}$.

Figure 5.17 indicates that the locus for constant input power operation of synchronous generator is a straight line perpendicular to the UPF line and its location with respect to point b (end of phasor V_t) determines the amount of power. So the no-load operating condition can be represented by a straight line perpendicular to the UPF line and passing through point b.

Figure 5.17 also indicates that variation of E_f (which is controlled by adjusting the field current) will affect the value of power factor angle θ and its lagging or leading status. These features can be easily recognized from Figure 5.17 which represents the operation of a synchronous machine whose armature resistance r_a is neglected and so $Z_s = jX_s$.

The circle diagram presented in Figure 5.18 reveals the following information:

1. The RL indicates the operation of machine. If the operating point is located above the RL, the machine is acting as a generator. But if the operating point is located below the RL, the machine is acting as a motor.

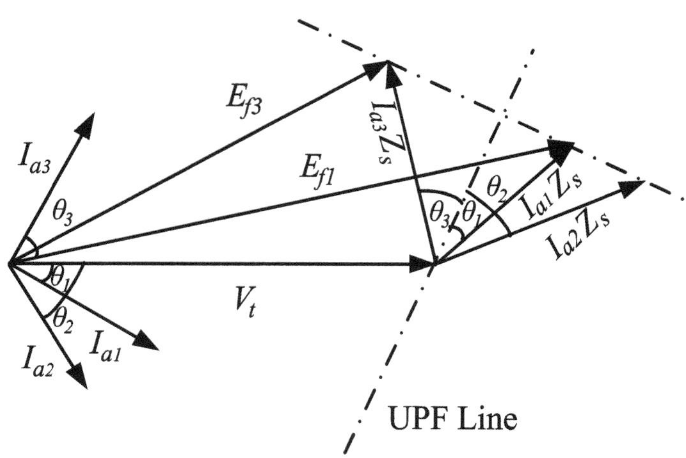

FIGURE 5.17 The locus for constant power operation of synchronous generator.

2. The UPF line indicates the power factor of the operating condition whether it is leading or lagging. For the generator case, increasing excitation (i.e., increasing E_f) will lead to lagging power factor and reducing excitation will result in leading power factor operation of the machine. The opposite is true for motor case.
3. The constant power line indicates that the steam admission to the steam turbine (i.e., the mechanical input power) is constant and so the input power is also constant.
4. The excitation circle represents fixed excitation for the machine field winding. This means a fixed generated electromotive force, E_f.
5. The steady state stability limit line indicates operation when the power angle δ equals 90°. The intersection of this intersection of this line and the excitation circle indicates the upper limit to power capability of the synchronous machine.

Example 5.7: A 3-phase, 80 kW, star-connected, 400 V synchronous alternator has asynchronous impedance per phase of $0.1 + j1\ \Omega$. If the DC excitation is equivalent to 440 line Vs, calculate the line current, power factor and efficiency when operating at full load, assuming no load losses of 2 kW.

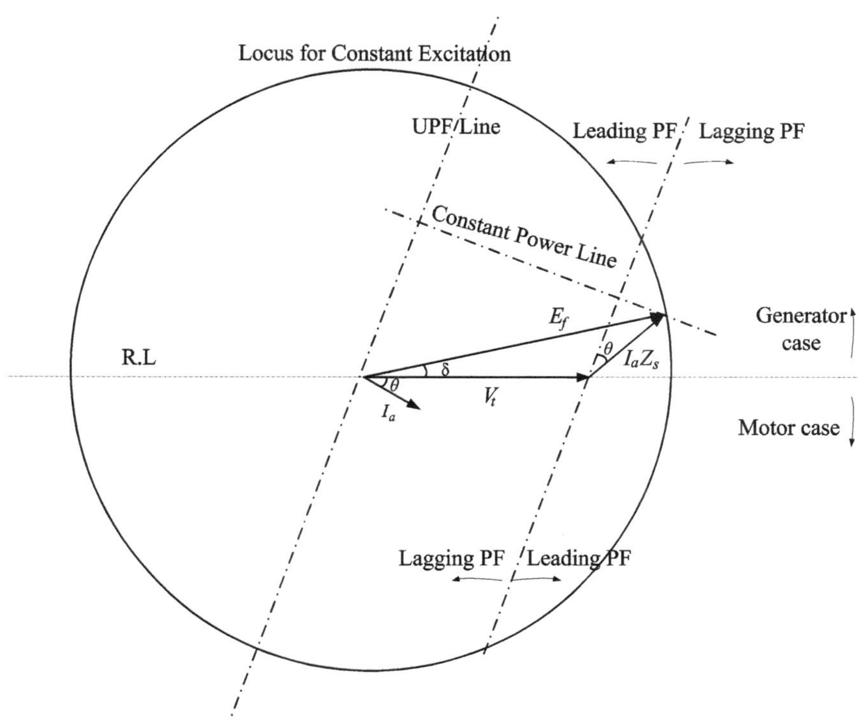

FIGURE 5.18 Circle diagram for the operation of synchronous machine.

Solution:

$$\alpha = \sin^{-1}\frac{r_a}{Z_s} = \sin^{-1}\frac{0.1}{\sqrt{0.1^2+1^2}} = \sin^{-1}\frac{0.1}{1.005} = 5.7^\circ$$

Using Equation (5.9)

$$80000 = \frac{(400/\sqrt{3})(400/\sqrt{3})^2}{1.005}\sin(\delta - 5.7) - \frac{(400/\sqrt{3})}{1.005}\cos(-5.7)$$

5.6 PARALLEL OPERATION AND LOAD DIVISION

It is rare to find a 3-phase alternator supplying its own load independently except under test conditions. In practice, a very large number of 3-phase alternators operate in parallel because the various power stations are interconnected through the national grid. Therefore, the output of any single alternator is small compared with the total interconnected capacity. The performance of a single alternator is unlikely to affect appreciably the voltage and frequency of the whole system. An alternator connected to such a system is said to be connected to infinite busbars. The outstanding electrical characteristics of such busbars are that they are constant-voltage, constant-frequency busbars.

There are many reasons for paralleling generators, such as:

1. For reliability of the power source;
2. When the load is reduced, one or more units can be shut down to keep the remaining units efficiently loaded;
3. Routine or emergency maintenance can be performed without disturbing the load conditions;
4. When added capacity is required, the new equipment can be simply paralleled with the old;
5. The equipment available to do a particular task may not be available in a sufficiently large capacity or size in a single unit.

5.6.1 Parallel Operation Requirements

There are a certain requirements that must be fulfilled for successful electrical paralleling. These are:

1. The terminal voltage (rms value) of the incoming alternator must be the same as bus bars voltage;
2. The frequency of the generated voltage of the incoming alternator must be equal to the bus bars frequency;
3. The phase of the incoming alternator voltage must be identical with the phase of the bus bars voltage. In other words, the two voltages must be in phase with each other;

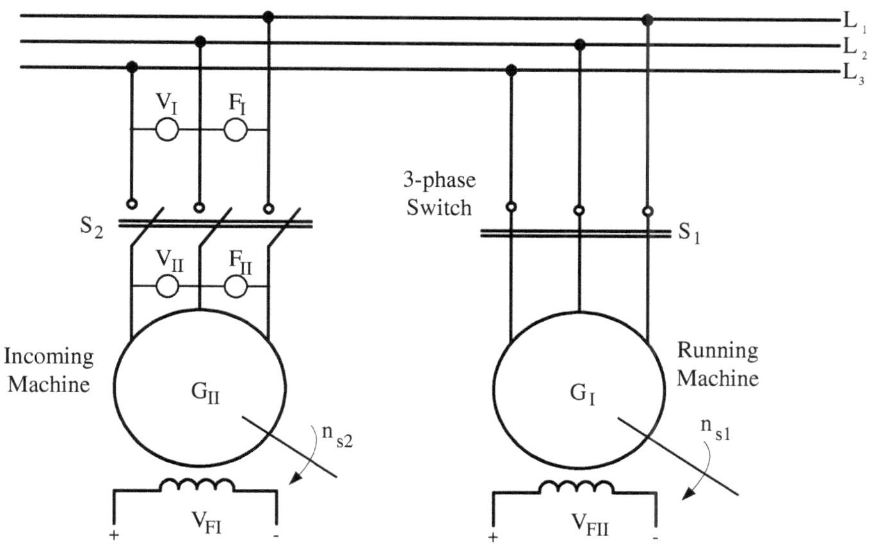

FIGURE 5.19 Parallel operation of synchronous generators

4. The phase sequence of the voltage of the incoming alternator should be the same as that of the busbars;
5. The prime movers must have similar speed-load characteristics. This is to prevent a machine with rising speed-load characteristics from taking more and more of the load unit it fails due to overloading.

Violation of these requirements for paralleling would result in circulating currents between the machine which may cause serious or even disastrous consequences. Figure 5.19 shows the connection diagram of two parallel generators.

5.6.2 Alternator Synchronizing Procedure

The method of connecting an incoming alternator safely to the live bus bars is called synchronizing. Two very simple and very reliable circuits show if phase sequence is correct. These are dark lamp circuit and the bright lamp circuit. They also serve as a means of testing for the in-phase condition. With the dark lamp circuit shown in Figure 5.20, the synchronizing switch open, and the incoming alternator not yet operating, the lamps will glow since they see the phase voltage and the full line frequency.

As the incoming alternator is brought up to speed, it has a voltage and frequency of its own. The lamps then see the difference of the voltage between the alternators. The desired phase sequence condition is known by whether the lamps go bright and dark together. If they twinkle on and off separately, there is a phase sequence error that must be corrected. The remedy is for any two of the three phase leads between

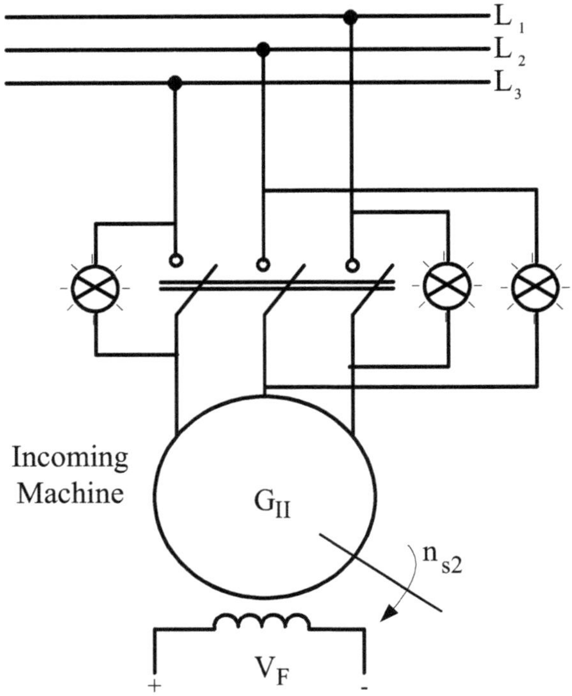

FIGURE 5.20 Synchronizing using dark lamp circuit.

the alternator and the switch to be interchanged. Finally, the lamps go dark and stay dark for a few seconds. The middle of the dark period is the point of actual in-phase condition. When this point has been identified satisfactorily, the synchronizing switch may be closed.

Figure 5.20 is the phasor diagram representing each machine, showing the line potential variations about a common natural potential.

The lamp voltage in the phasor diagram of Figure 5.21a is V_a, V_b, and V_c. The phasor voltages are sufficient to make the lamps glow. However, full brilliance comes when the two phasor sets are 180° degrees out of phase. The lamp voltage is then twice the line-to-neutral voltage. Figure 5.21b shows the ideal synchronizing situation, where there is no voltage acting on the lamp.

In large central stations, a device called a synchro scope is used for synchronization. A synchro scope has a rotating hand and a dial labeled with slow and fast direction arrows to show the incoming machine speed. In addition, an index point shows the actual in-phase position. During synchronization, as the incoming alternator rotational speed approaches near synchronism, the speed of the synchro scope hand drops enough to become visible. The hand speed is proportional to the difference in speed between the existing and incoming alternators. When the speeds are matched the hand will point to the index mark, whereupon the synchronizing switch may be closed. The final requirement for successful paralleling of ac machines is

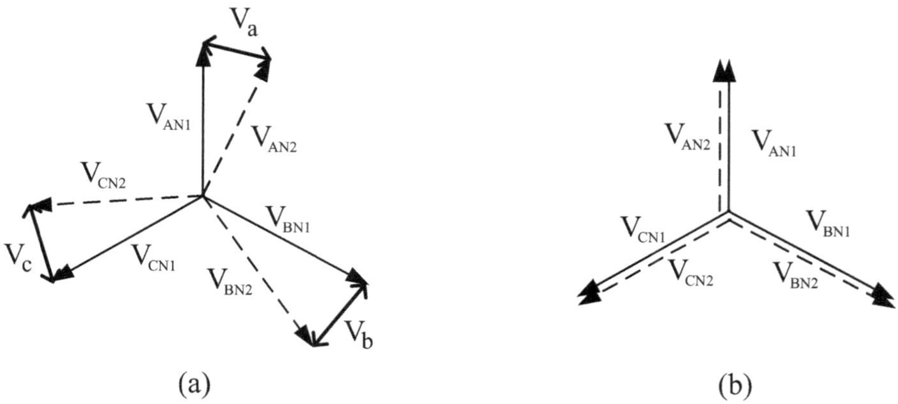

FIGURE 5.21 Dark lamp phasor diagram.

the speed-load characteristics of the prime mover. So, the alternators that are to be paralleled must have reasonably similar prime-mover speed-load characteristics and this speed-load characteristic must be drooping. This term means that as the torque load increases, the speed decreases. This condition allows the alternators to share the load equitably.

Example 5.8: A 3-phase, 1,000 kVA, 12 kV, wye-connected synchronous generator supplies 750 kW at 12 kV and 0.8 lagging power-factor load. The synchronous reactance is 30 Ω per phase, and the armature resistance is negligible. Calculate the voltage regulation.

SOLUTION:

(a) $V_t = \dfrac{12000}{\sqrt{3}} \angle 0^\circ = 6923.2 \angle 0^\circ \ V$

$$\therefore I_a = \frac{P_{out}}{\sqrt{3} V_L \cos\phi} \angle -\phi$$
$$= 45.1 \angle -36.87^\circ \text{A}$$

$$E_f = V_t + I_a(r_a + jX_s) = 6923.2\angle 0^\circ + (45.1 \angle -36.87^\circ)(j30)$$
$$= 7815.4\angle 8^\circ$$

$$\text{At no-load, } |V_{n.l}| = |E_f|$$

$$VR\% = \frac{|V_{n.l}| - |V_{fl}|}{|V_{fl}|} \times 100\% =$$
$$= \frac{7815.4 - 6923.2}{6923.2} = 12.816\%$$

Example 5.9: A 3-phase, 1,500 kVA, 13.2 kV, wye-connected synchronous generator has an armature resistance of 0.5 Ω and a synchronous reactance of 9.0 Ω. The generator supplying rated load at rated voltage. (a) Calculate the generated voltage at unity power factor, at 0.8 p.f. lagging, and at 0.8 leading power factor. (b) Calculate voltage regulation for each of the loads specified for part (a).

SOLUTION:

$$r_a = 0.5\,\Omega,\ X_s = 9.0\,\Omega$$

$$\text{(a) } V_t = \frac{13.2\times(10)^3}{\sqrt{3}}\angle 0^0 = 7621.02\ \angle 0^0\,\text{V}$$

$$\therefore I_{a.rated} = \frac{S_{rated}}{\sqrt{3}V_{L,rated}} = \frac{1500\times(10)^3}{\sqrt{3}\times 13.2(10)^3} = 65.6\text{ A}$$

At $pf = 1$

$$E_f = V_t + I_a(r_a + jX_s) = 7621.02\angle 0^\circ + (65.608\angle 0^\circ)(0.5 + j9)$$
$$= 7676.8\angle 4.41^\circ$$

$$VR\% = \frac{|V_{n.l}| - |V_{fl}|}{|V_{fl}|}\times 100\% =$$
$$= \frac{7676.6 - 7621.02}{7621.02}\times 100\% = 0.73\%$$

At $pf = 0.8$ lagging

$$E_f = V_t + I_a(r_a + jX_s) = 7621.02\angle 0^\circ + (65.608\ \angle -36.87^\circ)(0.5 + j9)$$
$$= 8014.3\angle 3.238^\circ$$

$$VR\% = \frac{|V_{n.l}| - |V_{f.l}|}{|V_{f.l}|}\times 100\% =$$
$$= \frac{8014.3 - 7621.02}{7621.02}\times 100\% = 5.16\%$$

At $pf = 0.8$ leading

$$E_f = V_t + I_a(r_a + jX_s) = 7621.02\angle 0^\circ + (65.608\ \angle 36.87^\circ)(0.5 + j9)$$
$$= 7306.6\angle 3.68^\circ$$

$$VR\% = \frac{|V_{nl}| - |V_{fl}|}{|V_{fl}|}\times 100\% =$$
$$= \frac{7306.6 - 7621.02}{7621.02}\times 100\% = -4.06\%$$

Example 5.10: A 3-phase, 6-pole, 60 Hz synchronous generator has a synchronous reactance of 4.0 Ω per phase and a terminal voltage of 2,300 V. The field current is adjusted so that the excitation voltage is 2,300 V at a power angle of 15°. Find

(a) Stator current
(b) Power factor
(c) Output power

SOLUTION:

$$r_a = 0.0, X_s = 4.0\,\Omega;\ \delta = 15^\circ\ \delta = 15^\circ$$

$$\text{(a)}\ E_f = \frac{12300}{\sqrt{3}}\angle 15^\circ = 1327.9\ \angle 15^\circ\ \text{V}$$

$$V_t = \frac{2300}{\sqrt{3}}\angle 0^\circ = 1327.9\ \angle 0^\circ\ \text{V}$$

$$\therefore I_a = \frac{E_f - V_t}{jX_s} = \frac{1327.9\ \angle 15^\circ - 1327.9\angle 0^\circ}{j4}$$

$$\therefore I_a = \frac{E_f - V_t}{jX_s} = \frac{1327.9\ \angle 15^\circ - 1327.9\ \angle 0^\circ}{j4} = 86.66\ \angle 7.15^\circ\ \text{A}$$

$$\text{(b)}\ pf = \cos(7.51^\circ) = 0.991\ \text{leading}$$

$$\text{(c)}\ \therefore P_{out} = 3\frac{E_f V_t}{X_s}\sin\delta = 3\frac{1327.9 \times 1327.9}{4}\sin 15$$

$$= 342.288\ kW$$

Example 5.11: A 200 kVA, 480 V, 60 Hz, 4-pole, Y-connected synchronous generator with a rated field current of 5 A was tested and the following data was taken.

a) from OC test – terminal voltage = 540 V at rated field current
b) from SC test – line current = 300 A at rated field current
c) from DC test – DC voltage of 10 V applied to two terminals, a current of 25 A was measured.

1. Calculate the speed of rotation in rpm
2. Calculate the saturated equivalent circuit parameters (armature resistance and synchronous reactance).

SOLUTION:

$$V_{rated} = \frac{480}{\sqrt{3}} = 277.13\ \text{V} \quad \text{and} \quad I_{a,\,rated} = \frac{200000}{\sqrt{3} \times 480} = 240.56\ \text{A}$$

$$Z_{base} = V_{rated} / I_{a,rated} = 277.13/240.56 = 1.152\ \Omega$$

The synchronous speed $N_s = 60\ f/P = 60 \times 60/2 = 3600/2 = 1800$ rpm

From DC Test

$$r_{DC} = 10/25 = 0.4\ \Omega$$
$$r_a = (0.4/2) \times 1.15 = 0.23\ \Omega$$
$$r_a = 0.23/1.152 = 0.2\ \text{PU}.$$

From OC Test

Rated field current produce rated voltage of $\frac{540}{\sqrt{3}} = 311.77$ V

From SC Test

Rated field current produce rated current = 300 A

$$\text{Saturated } Z_s = 311.77/300 = 1.04\ \Omega$$
$$\text{Saturated } Z_s = 1.04/1.152 = 0.903\ \text{PU}.$$

$$X_s = \sqrt{Z_s^2 - r_a^2}\sqrt{Z_s^2 - r_a^2} = \sqrt{1.04^2 - 0.23^2}\sqrt{1.04^2 - 0.23^2} = 1.0287\ \Omega$$
$$X_s = 0.893\ \text{PU}$$

Example 5.12: A 1,000 kVA, 3,300 V, 3-phase, star-connected alternator delivers full-load current at rated voltage at 0.80 p.f. lagging. The resistance and synchronous reactance of the machine per phase are 0.5 ohm and 5 ohms respectively.

Solution:

In case of lagging power factor

$$V_{rated} = \frac{3300}{\sqrt{3}} = 1905.26\ \text{V}$$
$$I_a = \frac{1000000}{\sqrt{3} \times 3300} = 174.95\ \text{A}$$
$$I_a r_a = 174.95 \times 0.5 = 87.475\ \text{V}$$
$$I_a X_S = 174.95 \times 5.0 = 874.75\ \text{V}$$

Component of E_f along the horizontal axis $= \text{V} + I_a\ ra\cos(\phi) + I_a X_S \sin(\phi) = 1905.26 + 87.475 \times 0.8 + 874.75 \times 0.6 = 2500.09$ V

Component of E_f along the vertical axis $= I_a\ X_S\cos(\phi) - I_a\ ra\sin(\phi) = 874.75 \times 0.8 - 87.475 \times 0.6 = 647.315$ V

$$E_f = \sqrt{2500.09^2 - 647.315^2} = 2582.53\ \text{V}$$
$$\text{Load angle } \delta = \tan^{-1}\frac{647.315}{2500.09} = 14.52$$

Example 5.13: The following test results were obtained on a 275 kW, 3-ph, 6,600 V non-salient pole type generator. Open-circuit characteristic:

Phase Voltage (V)	3233	3810	4180	4677
Field Current (A)	46.5	58	67.5	96

Short-circuit characteristic: Stator current 35 A with an exciting current of 50 A. Leakage reactance drop on full-load = 8%. Neglect armature resistance. Calculate as accurately as possible the exciting current (for full-load) at power factor 0.8 lagging.

SOLUTION:

$$V_t = \frac{6000}{\sqrt{3}} = 3810.5 \text{ V}$$

$$I_a = \frac{275000}{\sqrt{3} \times 6600} = 30 \text{ A}$$

$$I_a\, X_l = 8\% = 0.08 \text{ P.U.} = 0.08 \times 3810.5 \text{ V.}$$

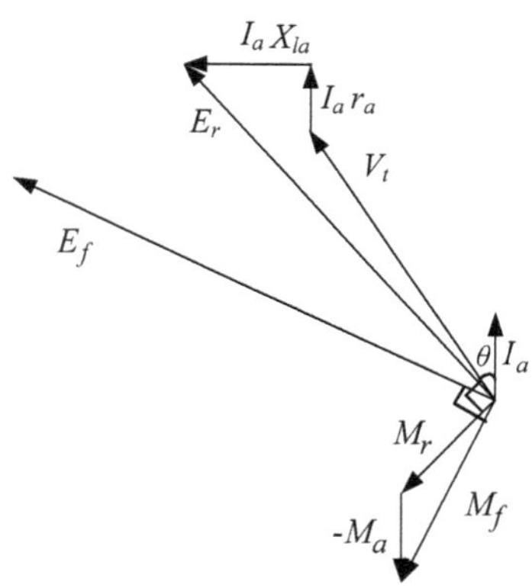

From the phasor diagram shown in the figure, draw the rated voltage as a reference. Then draw $I_a X_{la} = 308.4 \text{ V} \perp I_a$. The resultant voltage E_r can be obtained graphically by the length = 4,010 V.

From the OCC, at voltage of 4,010, we can get $M_r = 62 \text{ A}$. and represented by $M_r \perp E_r$.

From the SCC, at full load current of 30 A, the corresponding field current $M_a = 43$ A that is represented by the line // I_a and in the same direction. From the MMF triangle, the value of M_f that is equal to 94.3A.

Example 5.14: A 3-phase, 6,000 V alternator has the following OCC at normal speed:

Field Current (A)	14	18	23	30	43
Terminal Voltage (V)	4000	5000	6000	7000	8000

With armature short-circuited and full-load current flowing, the field current is 17 A.

Also when the machine is supplying full-load of 2,000 kVA at zero power factor, the field current is 42.5 A and the terminal voltage is 6,000 V.

Determine the field current required when the machine is supplying the full-load at 0.8 p.f. lagging. The OCC at phase voltage is

Field Current (A)	14	18	23	30	43
Phase Voltage (V)	2309.4	2886.7	3464.1	4041.4	4618.8
Field Current (A)	17	42.5			
Phase Voltage (V)	0	3464.1			

SOLUTION:

From both the OCC and ZPFC, we can obtain Potier triangle BHD as shown in the figure below.

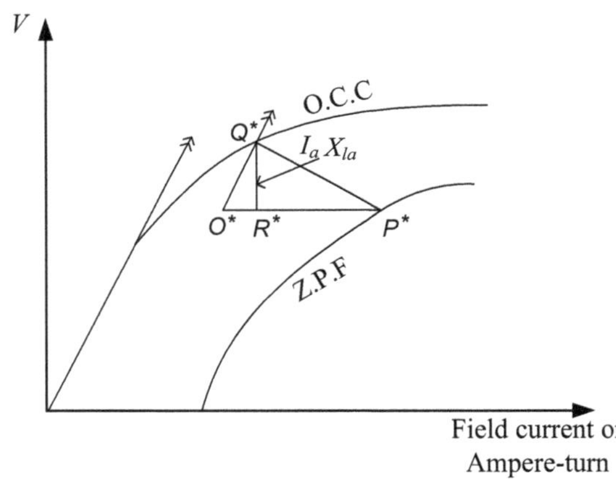

From Potier triangle, Q×R× represents $I_a X_{la} = 450$ V.

From the phasor diagram, we can obtain E_r graphically or mathematically as:

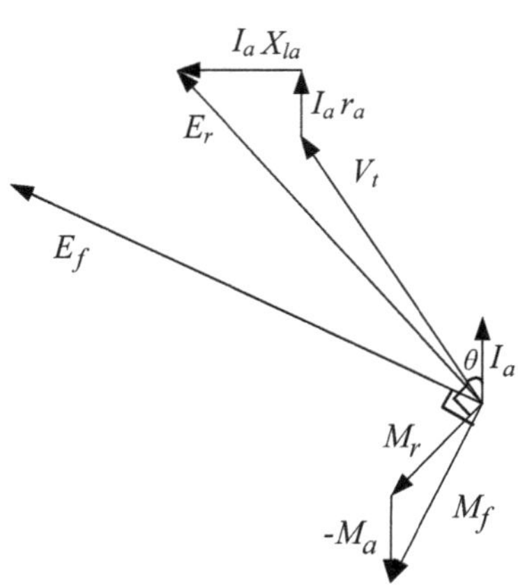

$$E_r = \sqrt{(V_t + I_a X_{La}\sin\phi)^2 - (I_a X_{la}\cos\phi)^2}$$
$$= \sqrt{(3464.1 + 50\times 0.6)^2 - (450\times 0.8)^2} = 3751.41\ \text{V}$$

From the OCC at 3,751 V, we get the corresponding field current (M_r) of 26.5 A. This field current (26.5) A is drawn $\perp E_r$.

From Potier triangle we can get the field current (M_a corresponding to $I_a X_{al}$) which is equal to 14.5 A. This value is drawn parallel to the armature current and in the same direction.

Therefore, from the MMF triangle, the field current $(M_f) = 37.2$ A

Example 5.15: An 11 kV, 1,000 kVA, 3-phase, Y connected alternator has a resistance of 2 Ω per phase. The open-circuit and full-load zero power factor characteristics are given below. Find the voltage regulation of the alternator for full load current at 0.8 p.f. lagging by Potier method.

Field Current (A)	40	50	110	140	180
OCC line voltage (V)	5800	7000	12500	13750	15000
ZPF line voltage (V)	0	1500	8500	10500	12500

SOLUTION:

The phase voltages used in OCC and ZPFC are as in the following table:

Field Current (A)	40	50	110	140	180
OCC line voltage (V)	3348.63	4041.45	7216.88	7938.5	8660.2
ZPF line voltage (V)	0	866	4907.48	6062.2	7216.9

$$V_t = \frac{11000}{\sqrt{3}} = 6350.85\ \text{V}$$

$$I_a = \frac{1000000}{\sqrt{3}\times 11000} = 52.49\ \text{A}$$

Potier triangle is drawn as shown in the figure below. Where $Q^*P^* = 40$ A. O^*Q^* is drawn parallel to the air gap line. Q*R* ⊥ O*P*.

Q*R* represents the leakage reactance drop $I_a X_{la} = 1000$ V (by measurement).

$R^*P^* = 30$ A (armature reaction M_a)

We draw the phasor diagram of the alternator at 0.8 p.f. lag as shown in the figure below.

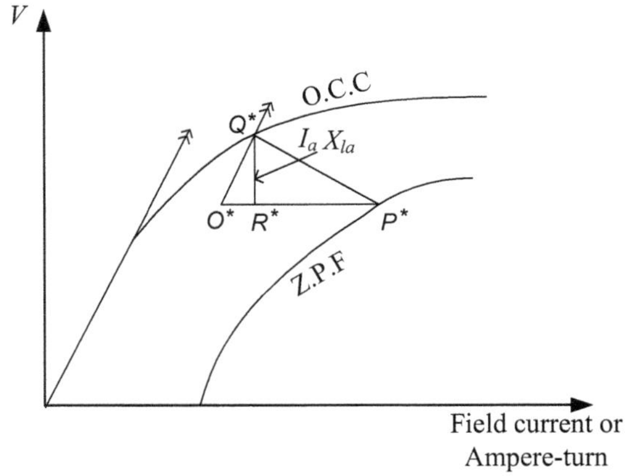

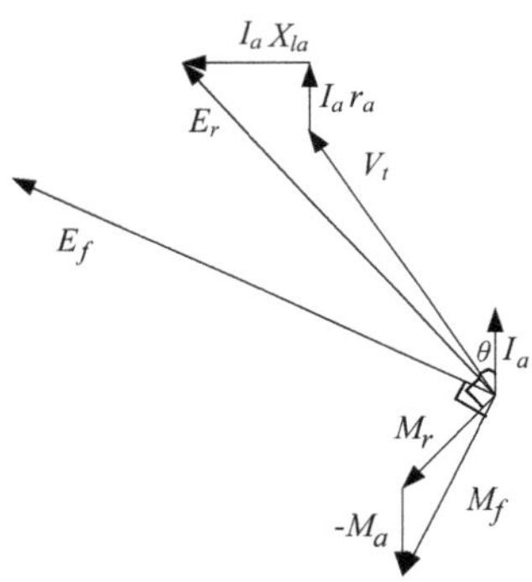

Draw $I_a r_a = 52.49 \times 2 = 105$ V represented by the length $// I_a$

Then draw $I_a X_{la} = 1000$ V represented by the length $\perp I_a$

Graphically, we can obtain the air gap voltage $E_r = 7080$ V

From OCC at 7,080 V, the corresponding field current (M_r) is 108 A that is drawn $\perp E_r$.

From Potier triangle $M_a = 30$ A and represented by the line // I_a and in the same direction.

From the MMF triangle, the length that represent M_f is obtained = 128 A.

From the OCC at field current = 128, the corresponding voltage $(E_f) = 7700$ V.

$$VR\% = \frac{|V_{n.l}| - |V_{f.l}|}{|V_{f.l}|} \times 100\%$$

$$= \frac{7700 - 6350}{6350} \times 100\% = 0.2126 = 21.26\%$$

Problems

5.1 The number slots of a 3-phase, 12-pole, Y-connected alternator has 180. The number of conductor per slot is 12 and the conductors are connected in series. The coil is full-pitch winding and the flux per pole is 0.05 Wb. If the machine runs at a speed of 600 rpm, determine the line voltage.

5.2 A 3-phase, 50 Hz, 4-pole, Y-connected alternator has six slots per phase. The armature has double layers winding. The windings are arranged as five conductors per layer. The coil pitch is 5/6th of the full pitch and the flux per pole is 0.05 Wb. If the machine runs at a speed of 600 rpm, determine the phase voltage.

5.3 A 3-Φ, 50 Hz, 10-pole, synchronous generator has three slots/phase. The armature has double layers winding. The windings are arranged as five conductors/layer. The coil span is 150°. The value of the fundamental and third harmonic component of the flux is 0.05 Wb and 0.006 Wb. Determine the rms value of the induced voltage per phase.

5.4 The number slots of a 3-Φ, 4-pole, Y-connected alternator has 36 slots. The number of conductor per slot is eight and the conductors are connected in series. The coil is full pitch winding and the flux per pole is 0.06 Wb. If the machine runs at a speed of 600 rpm, determine the line voltage.

5.5 The number slots/pole/phase of a 3-Φ, 50 Hz, 4-pole, Y-connected alternator is four. The armature has double layers winding. The windings are arranged as 4 conductors/layer. The coil pitch is 3/4th of the full pitch and the flux/pole is 0.03 Wb. Determine the phase and line voltage.

5.6 A 3-Φ, 60 Hz, 6-pole, synchronous generator has two slots/phase. The armature has double layers winding. The windings are arranged as four conductors/layer. The coil span is 150°. The value of the fundamental of the flux is 0.05 Wb and third harmonic component of the flux is 30% of the fundamental flux. Determine the rms value of the induced voltage/phase.

5.7 Two alternators connected in parallel to deliver electrical power of load 5 Ω. The necessary parameters are: $E_1 = 220\text{V}$, $Z_1 = j3\ \Omega$; $E_2 = 220\ \ 5°\ \text{V}$, $Z_2 = j4\ \Omega$. Determine the load current, terminal voltage, power/phase delivered by the first alternator.

5.8 Two alternators connected in parallel to deliver electrical power of load 8 Ω. The necessary parameters are: $E_1 = 210\text{V}$, $Z_1 = j5\ \Omega$; $E_2 = 210\ \ 5°\ \text{V}$, $Z_2 = j6\ \Omega$. Find the load current, terminal voltage and 3-Φ Power delivered by the second alternator.

5.9 In a 50 kVA, Y-connected, 440 V, 3-phase, 50 Hz alternator, the effective armature resistance is 0.25 Ω per phase. The synchronous reactance is 3.2 Ω per phase and leakage reactance is 0.5 Ω per phase. Determine at rated load and unity power factor:

(a) Line value of the internal resultant EMF, E_r
(b) Line value of the no-load EMF, E_f
(c) Percentage regulation on full-load
(d) Armature reactance.

5.10 A 1,000 kVA, 3,300 V, 3-phase, Y-connected alternator delivers full-load current at rated voltage at 0.80 p.f. lagging. The resistance and synchronous reactance of the machine per phase are 0.5 Ω and 5 Ω, respectively. Estimate the terminal voltage for the same excitation and same load current at 0.80 p.f. leading.

5.11 Find the synchronous impedance and reactance of an alternator in which a given field current produces an armature current of 200 A on short-circuit and a generated EMF of 50 V on open-circuit. The armature resistance is 0.1 Ω. To what induced voltage must the alternator be excited if it is to deliver a load of 100 A at a p.f. of 0.8 lagging, with a terminal voltage of 200V?

5.12 From the following test results, determine the voltage regulation of a 2,000 V (phase), Y-connected 3-phase alternator delivering a current of 100 A at (i) unity p.f. (ii) 0.8 leading p.f. and (iii) 0.71 lagging p.f.

Test results: Full-load current of 100 A is produced on short-circuit by a field excitation of 2.5A. An EMF of 500 V is produced on open-circuit by the same excitation. The armature resistance is 0.8 Ω.

5.13 A 100 kVA, 3,000 V, 50 Hz, 3-phase, star-connected alternator has effective armature resistance of 0.2 Ω. The field current of 40 A produces short-circuit current of 200 A and an open-circuit EMF of 1,040 V (line value). Calculate the full-load voltage regulation at 0.8 p.f. lagging and 0.8 p.f. leading. Draw phasor diagrams.

5.14 A 3-phase, star-connected alternator is rated at 1,600 kVA, 13,500 V. The armature resistance and synchronous reactance are 1.5 Ω and 30 Ω, respectively, per phase. Calculate the percentage regulation for a load of 1,280 kW at 0.8 leading power factor.

5.15 A 3-phase, 10 kVA, 400 V, 50 Hz, Y-connected alternator supplies the rated load at 0.8 p.f. lag. If the resistance is 0.5 Ω and syn. reactance is 10 Ω, find the power angle and voltage regulation.

5.16 The following test results are obtained from a 3-phase, 6,000-kVA, 6,600 V, Y-connected, 2-pole, 50 Hz turbo-alternator:

With a field current of 125 A, the OC voltage is 8,000 V at the rated speed; with the same field current and rated speed, the SC current is 800 A. At the rated full-load, the resistance drop is 3%. Find the regulation of the alternator on full-load and at a power factor of 0.8 lagging.

5.17 A 3-phase, Y-connected, 3,300 V, synchronous generator which supplies 1,000 kW at 80% power factor lag, if its OCC is given by the following data:

Field Current (A)	80	96	118
OCC line voltage (V)	3300	3600	3900

the stator winding has a per phase resistance of 0.15 Ω and a leakage reactance of 1.2 Ω. There are 16 poles, each wound with 108 turns. The armature winding is arranged in 144 slots, 5 conductors per slot, with full-pitched coils.

a) Calculate the armature MMF in AT/Pole

b) Estimate the field current (I_f) for this loading condition

5.18 Calculate the saturated synchronous reactance (in Ω/phase and per unit) of a 85 kVA synchronous machine which achieves its rated open-circuit voltage of 460 V at a field current 8.7 A and which achieves rated short-circuit current at a field current of 11.2 A.

5.19 A synchronous generator with 40% synchronous reactance and negligible resistance is supplying 3/4 of the full load current at 0.7 p.f. lag and at normal terminal voltage. If the current rises to the full load value at 0.6 p.f. lag, determine the percentage change in the terminal voltage if the excitation is kept constant.

5.20 A six-pole round rotor, 3-phase Y-connected synchronous machine has the following tests:

Open circuit test: 4000 V line to line at 1,000 rpm, 50 A rotor current

Short circuit test: 300 A at 500 rev/min 50 A rotor current

Neglect the stator resistance and core losses, and assuming a linear open circuit characteristic, calculate:

(a) The machine synchronous reactance at 50 Hz,

(b) The rotor current required for the machine to operate as a motor at 0.8 power factor leading from a supply of 3.3 kV line to line with an output power of 1,000 kW,

(c) The rotor current required for the machine to operate as a generator on an infinite bus of 3.3 kV line to line when delivering 1,500 kVA at 0.8 power factor lagging,

(d) The load angle for (b) and (c) and sketch the phasor diagram for (b) and (c).

5.21 A 3-phase, 8-pole, 50 Hz, 6,600 V, Y-connected synchronous motor has a synchronous impedance of $0.66 + j6.5\ \Omega/\text{phase}$. When excited to give EMF of 4,500 V/phase, it takes an input power of 2,500 kW. Calculate the electromagnetic torque, input current, p.f., and the load angle.

5.22 For a 3-phase, Y-connected 2,500 kVA, 6,600 V, synchronous generator operating at full load, calculate:

(a) The percentage of voltage regulation at a power factor of 0.8 lagging,

(b) The percentage of voltage regulation at a power factor of 0.8 leading.

The synchronous reactance and the armature resistance are 10.4 Ω and 0.071 Ω, respectively.

5.23 A Y-connected, 3-phase, 50 Hz, 8-pole, synchronous alternator has an induced voltage of 4,400 V between the lines when the rotor field current is 10 A. If this alternator is to generate 60 Hz voltage, compute the new synchronous speed and induced voltage for the same rotor current of 10 A.

5.24 A 3-phase, Y-connected, six-pole alternator is rated at 10 kVA, 220 V, at 60 Hz. Synchronous reactance $X_s = 3\ \Omega$. The no load line to neutral terminal voltage at 1,000 rpm follows the magnetization curve is as follows:

E (V)	11	38	70	102	131	156	178	193	206
I_f (A)	0	0.2	0.4	0.6	0.8	1.0	1.2	1.4	1.6
E (V)	215	221	224						
I_f (A)	1.8	2.0	2.2						

Determine:

(a) The rated speed in rpm,

(b) The field current required for full load operation at 0.8 power factor lagging.

5.25 The following values give the open-circuit and full-load zero p.f. saturation curves for a 15,000 kVA. 11,000 V, 3-phase, 50 Hz, star-connected turbo-alternator:

Field AT	1000:	10	18	24	30	40	45	50
OC line in (kV):	4.9	8.4	10.1	11.5	12.8	13.3	13.65	
Zero p.f. in (kV):	—	0	—	—	—	10.2	—	

Find the armature reaction, the armature reactance and the synchronous reactance.

Deduce the regulation for full-load at 0.8 power lagging.

5.26 A 600 kVA, 3,300 V, 8-pole, 3-phase, 50 Hz alternator has following characteristic:

Amp-turns/pole:	4000	5000	7000	10,000	
Terminal EMF:		2850	3400	3850	4400

There are 200 conductors in series per phase.

Find the SCC, the field ampere-turns for full-load 0.8 p.f. (lagging) and the voltage regulation.

Given that the inductive drop at full-load is 7% and that the equivalent armature reaction in amp-turns per pole = 1.06 × ampere-conductors per phase per pole.

5.27 The following data are taken from the open circuit and short circuit characteristics of a 45 kVA, 3-phase, Y-connected, 220 V (line to line), 6-pole, 60 Hz synchronous machine:

OCC: Line to line voltage = 220 V Field current = 2.84 A

SCC: Armature current (A) 118 152 Field current (A) 2.20 2.84

Air gap line: Field current = 2.20 A line to line voltage = 202 V

Calculate the unsaturated value of the synchronous reactance, and its saturated value at the rated voltage. Express the synchronous reactance in ohms per phase and also in per unit on the machine rating as a base.

5.28 The open-and short-circuit test readings for a 3-phase, star-connected, 1,000 kVA, 2,000 V, 50 Hz, synchronous generator are:

Field Current (A)	10	20	25	30	40	50
OC line voltage (V)	800	1500	1760	2000	2350	2600
SC current (A)	--	200	250	300	--	--

The armature effective resistance and leakage reactance are 0.2 Ω and 0.6 Ω per phase, respectively. Draw the characteristic curves and estimate the full-load percentage regulation at 0.8 p.f. lag.

6 Synchronous Motor

6.1 INTRODUCTION

The construction of the synchronous motor is the same as the that of a synchronous generator. As with a generator, a DC generator may be mounted on the motor shaft to excite the field. However, in the most common type of field construction, the poles are somewhat different from those used in the generator. In addition to the exciting windings, the synchronous motor poles are required with a squirrel-cage windings. The slots of this cage windings are located in the pole faces and are parallel to the shaft axis. The terminals of the copper bars embedded in these slots are short circuited in the same manner as in a squirrel-cage. This short-circuited winding on the rotor is known as the damper windings or the starting windings. These additional windings will help the motor to start from a standstill condition and will damp any hunting developed.

For a synchronous motor, motor action results from the interaction between the DC magnetic field which is developed by the rotating poles and the field created by the 3-phase currents flowing through the uniformly distributed 3-phase stator windings. Thus, there will be two rotating magnetic fields available in the airgap. One is rotating at speed equal to that of the rotor, and the second which is the resultant 3-phase resolving field, revolves at synchronous speed.

So, the synchronous motor (whose name means a motor that rotates in synchronism with its revolving field developed by its stator windings) will operate at a speed which is determined from the number of poles and the supply frequency. The synchronous motor action reveals that the speed regulation of synchronous motor at any load torque is always 0%.

6.2 EFFECT OF INCREASING OR DECREASING THE EXCITATION

Intuitively, we can expect that increasing the strength of the magnets will increase the magnetic attraction, and thereby cause the rotor magnets to have a closer alignment with the corresponding opposite poles of the rotating magnetic pole s of the stator. This will obviously result in a smaller power angle. When the shaft load is assumed to be constant, the steady-state value of $E_f \sin\delta$ must also be constant. An increase in E_f will cause a transient increase in $E_f \sin\delta$, and the rotor will accelerate. As the rotor changes its angular position, δ decreases until $E_f \sin\delta$ has the value as before, at which time the rotor is again operating at synchronous speed, as it should run only at the synchronous speed. This change in angular position of the rotor magnets relative to the poles of rotating magnetic field of the stator occurs in a fraction of a second. The effect of changes in field excitation on armature current, power angle, and power factor of a synchronous motor operating with a constant shaft load, from a constant voltage, constant frequency supply, is illustrated in Figure 6.1. For a constant shaft load:

DOI: 10.1201/ 9781003795841-6

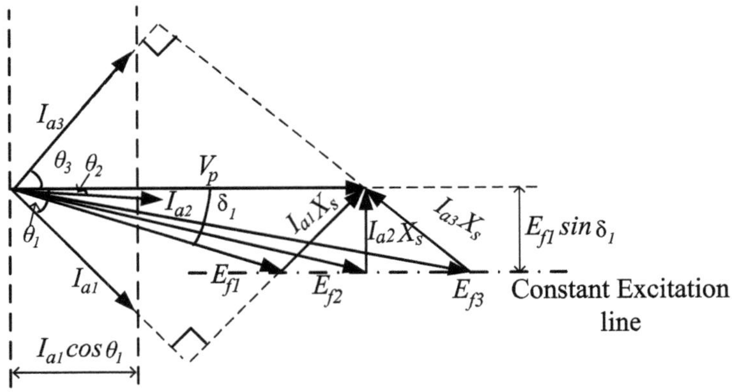

FIGURE 6.1 Phasor diagram showing effect of changes in field excitation on armature current, power angle and power factor of synchronous motor.

$$E_{f1}\sin\delta_1 = E_{f2}\sin\delta_2 = E_{f3}\sin\delta_3 = E_f\sin\delta \tag{6.1}$$

This is shown in Figure 6.2, where the locus of the tip of the E_f phasor is a straight line parallel to the V_p phasor. Similarly, for a constant shaft load:

$$I_{a1}\cos\phi_1 = I_{a2}\cos\phi_2 = I_{a3}\cos\phi_3 = I_a\cos\phi \tag{6.2}$$

This is also shown in Figure 6.1, where the locus of the tip of the I_a phasor is a line perpendicular to the V_p phasor.

To summarize the previous discussion, a change in excitation away from the normal excitation will result the following:

1. A change in armature current, and
2. A corresponding change in the power factor of the motor.

6.3 EFFECT OF INCREASING MECHANICAL LOAD

It must be realized that increasing the mechanical load connected to the shaft of the synchronous motor, increases the torque angle since the motor always rotates at synchronous speed. If the DC excitation is kept constant so that the flux remains the same throughout the operation range, the armature current will increase because of the increase in the resultant EMF, E_r, as shown in Figure 6.2.

Although the armature current increases as the load increases, both the developed power and torque will not increase indefinitely. It can be shown analytically that the developed power and torque increase with load until the torque angle is equal to the synchronous impedance angle β = ($\beta = \tan^{-1}\frac{X_s}{r_a}$). At this point, a further increase in load will cause the motor to pull out of synchronism.

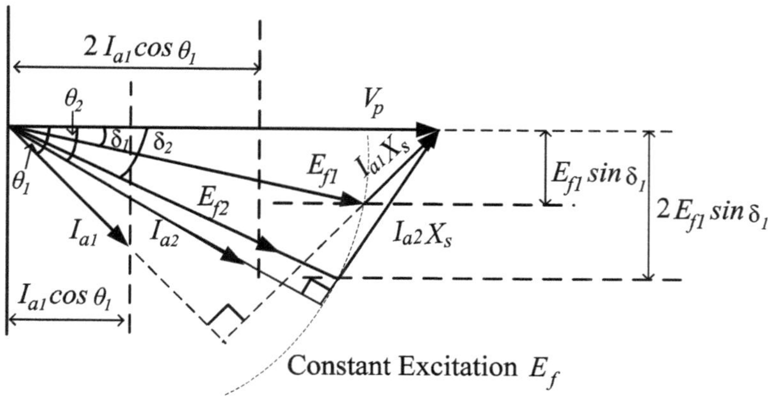

FIGURE 6.2 Phasor diagram showing effect of changes in shaft load on armature current, power angle and power factor of a synchronous motor.

Example 6.1: A 10 hp, 415 V, 50 Hz, 4-pole, Y-connected synchronous motor is operating with a light load. The corresponding power (torque) angle is 3°. The DC excitation is adjusted for a counter EMF per phase of 210 V.

(a) What is the resultant armature voltage per phase?
(b) What is the angle between E_r and V_p?

SOLUTION

$$V_p = \frac{415}{\sqrt{3}} = 240 \text{ V}$$

Since $E_c < V_P$, the motor is operating at the under-excitation condition, and the corresponding phasor diagram will be:

(a) $$E_r = \sqrt{(V_p - E_c\cos\alpha)^2 + (E_c\sin\alpha)^2}$$
$$= \sqrt{(240 - 210\times 0.999)^2 + (210\times 0.052)^2} = 32 \text{ V}$$

(b) $$\alpha = \arctan\frac{E_c\sin\delta}{V_p - E_c\cos\delta} = \arctan\frac{210\times 0.052}{240 - 210\times 0.999} = 19.8^0$$

Example 6.2: The motor of example 6.1 has its DC excitation adjusted to give a counter EMF per phase of 240 V. For the same angular displacement of the rotor, find

(a) the resultant armature voltage;
(b) the angle between E_r and V_p.

SOLUTION

Since $E_c = V_p$, the motor is operating at the normal excitation condition, and the corresponding phasor diagram will be:

since $\delta = 3^0$

$$E_r = \sqrt{(240 - 240 \times \cos 3)^2 + (240 \times \sin 3)^2} = 12.48 \text{ V}$$

$$\alpha = \arctan \frac{240 \times \sin 3}{240 - 240 \times \cos 3} = 89.97^0$$

Example 6.3: A 25 hp, 415 V, 50 Hz, 4-pole, Y-connected synchronous motor is driving a mechanical load. The corresponding torque angle is 10°. The motor has an armature resistance per phase of 0.24 ohm and a synchronous reactance per phase of 2.6 ohms.

If the excitation is adjusted for a counter EMF per phase of 215 V, find (a) the armature current; (b) the power factor angle; (c) the total power to the motor. Repeat for the case when the DC excitation is readjusted for a counter EMF per phase of 240 V.

SOLUTION:

(i) The case when $E_c = 215$ V (under-excitation condition)
The corresponding phasor diagram will be:

$$E_r = \sqrt{(V_p - E_c \cos\alpha)^2 + (E_c \sin\alpha)^2}$$

$$= \sqrt{(240 - 215 \times 0.985)^2 + (210 \times 0.1736)^2} = 46 \text{ V}$$

$$\alpha = \arctan \frac{E_c \sin\delta}{V_p - E_c \cos\delta} = \arctan \frac{37.324}{28.225} = 52.9^0$$

$$Z_s = \sqrt{(X_s)^2 + (r_s)^2} = \sqrt{(0.24)^2 + (2.6)^2} = 2.61 \text{ ohms}$$

(a) $I_a = \dfrac{E_r}{Z_s} = \dfrac{46}{2.61} = 17.6$ A

$$\beta = \arctan \frac{X_s}{r_a} = 84.7^0$$

(b) $\therefore \theta = \beta - \alpha = 84.7 - 52.9 = 31.8^0$

$\therefore \cos\theta = 0.85$ lagging

(c) The input power $= 3\, V_p I_a \cos\theta = 2 \times 240 \times 17.6 \times 0.85$
$= 10.77$ kW

(ii) The case when $E_c = 240$ V(Normal-excitation condition)

$$E_r = 2V_p \cos\left(\frac{180-\delta}{2}\right) = 2\times 240\times \cos 85 = 41.8 \text{ V}$$

For normal-excitation Er bisect the angle between E_c and V_p.

$$\alpha = \frac{1}{2}(180-\delta)$$

(a) $I_\alpha = \dfrac{E_r}{Z_s} = \dfrac{41.8}{2.61} = 16$ A

(b) $\therefore \theta = \beta - \alpha = 84.7 - 85 = 0.3^0$

$\therefore \cos\theta = 0.99$ leading

(c) The input power $= 3\ V_p I_a \cos\ \theta = 3\times 240\times 16\times 0.9$

$= 11.4$ kW

It is obvious that the active power has been increased although the reactive component of the power is decreased.

Example 6.4: Consider a 125 hp, 2,300 V, 50 Hz, 4-pole, Y-connected synchronous motor. The armature resistance per phase of 0.8 ohm and a synchronous reactance per phase of 16 ohms. When operated at no-load, the power input is 6,000 W and the armature current is 5 A. If the motor is operating at a torque (power) angle of 15° and the counter EMF is 1,500 V per phase (over-excited), calculate: (a) the power output in horsepower; (b) the output torque and (c) efficiency.

If the mechanical load is increased until the torque (power) angle is equal to the synchronous impedance angle, recalculate (a), (b) and (c), and then find the ratio of the corresponding values.

Solution

By using the no-load values:

The rotational losses $P_{rot} = P_{in} - 3I_a^2 r_a$

$= 6000 - 3\times 5^2 \times 0.8$

$= 5940$ W

The rotational losses are considered constants, since the flux remains the same throughout the operating range (i.e., the core losses are constant) and the motor speed is always constant (i.e., fractional losses are constant).

(a) $E_r = \sqrt{(V_p - E_c\cos\delta)^2 + (E_c\sin\delta)^2}$

$= \sqrt{(1330 - 1500\times 0.966)^2 + (1500\times 0.259)^2} = 406$ V

$$\alpha = \tan^{-1}\frac{388.5}{-119} = 107^0$$

$$I_a = \frac{E_r}{Z_s} = \frac{406}{\sqrt{(0.8)^2 + (16)^2}} = 25.37 \text{ A}$$

$$\beta = \arctan\frac{X_s}{r_a} = 87^0$$

$$\therefore \theta = \beta - \alpha = 87 - 107 = 20^0 \text{ leading}$$

$$\therefore \cos\theta = 0.94 \text{ leading}.$$

The electrical power developed = The input power – armature copper losses

$$P_d = 3V_p I_a \cos\theta \text{-} 3I_a^2 r_a = 3\times1330\times25.3\times0.94 \text{ - } 3$$
$$\times\, 25.37^2 \times 0.8 = 93.608 \text{ kW}$$

The output power = P_d - rotational losses = 87.668 kW
=117.5 hp

(b) The output torque $= \dfrac{P_o}{2\pi\dfrac{S_s}{60}} = 558$ Nm

(c) The efficiency $\xi = \dfrac{87.668}{2\times1330*25.37*0.94} * 100 = 92.1$ %

Now, at pull-out conditions

$$\delta = \beta = 87^0$$

(a) $E_r^* = \sqrt{(1330 - 1500\times\cos 87)^2 + (1500\times\sin 87)^2}$
= 1952 V

$$\alpha^* = \tan^{-1}\frac{1498}{1251.5} = 50^0$$

$$I_a^* = \frac{E_r^*}{Z_s} = 122 \text{ A}$$

$$\therefore \theta^* = \beta - \alpha^* = 87 - 50 = 37^0 \text{ leading}$$

$$\therefore \cos\theta^* = 0.798 \text{ leading}.$$

The electrical power developed = $3\times1330\times122\times0.798 - 3$
$\times 122^2 \times 0.8 = 352.729$ kW

The output power $P_o^* = 352.729$ kW-5940 = 346.789 kW
= 464.8 hp

(b) The output torque $T_o^* = \frac{P_o^*}{2\pi \frac{S_s}{60}} = 2207.7 \text{ Nm}$

(c) The efficiency $\xi^* = \frac{346.789}{2*1330*122*0.798} * 100 = 89.27\ \%$

THE RATIOS:

$$\frac{P_o^*}{P_o} = \frac{T_o^*}{T_o} = 3.95$$

$$\frac{\xi^*}{\xi} = 0.969$$

6.3.1 SYNCHRONOUS MOTOR STARTING

The theory of torque production by a synchronous motor requires that the rotor rotates at the same speed as the stator resultant field. It is known that the 3-phase currents in the armature conductors of the stator produce a uniform magnetic field rotating the synchronous speed. If the rotor is going at some other speed, the equivalent rotating stator-field poles will be moving past the rotor poles, first attracting, then repelling them. The average torque under these conditions is zero. As a result, if the rotor and stator of a standstill synchronous motor are excited, the motor will develop no torque and so it will not start by itself. It is evident, then, that the synchronous motor must brought up to a speed sufficiently close its synchronous speed, in order to lock into synchronism with the stator rotating field. There are three methods that are used to start a synchronous motor:

(1) Use an external prime mover to accelerate the rotor of synchronous motor near to its synchronous speed and then supply the rotor as well as stator. Of course, care should be taken to ensure that the direction of rotation of the rotor as well as that of the rotating magnetic field of the stator are the same. This method is usually followed in the laboratory—the synchronous machine is started as a generator and is then connected to the supply mains by following the synchronization or paralleling procedure. Then the power supply to the prime mover is disconnected so that the synchronous machine will continue to operate as a motor.
(2) Reduce the speed of the rotating magnetic field of the stator to a low enough value that the rotor can easily accelerate and lock in with it during one half-cycle of the rotating magnetic field's rotation. This is done by reducing the frequency of the applied electric power.
(3) The most widely used means of starting is achieved through the use of damper windings (the starting windings) which consist of heavy copper bars installed in slots in the pole faces of synchronous motor. These bars

which are shorted together at both ends of the rotor, will form the rotor during starting, to act as a squirrel-cage rotor. So, at starting, currents are induced are induced in the short-circuited bars by the rotating air gap field which is created from the 3-phase currents passing through the armature conductors. These induced currents interact with the air gap field to produce a torque in the direction of field rotation. In other words, the machine is started as an induction motor. Induction motor action will bring the machine to nearly synchronous speed and so the DC field current is turned on. The heavy bar windings in the pole faces serve an additional purpose. When there is any change in load, this results in an oscillatory motion superimposed on the normal synchronous speed of the motor. This action is called hunting. The hunting effect will induce currents in the heavy bar windings. The field set up by these currents produced a torque that opposes any change in torque angle δ, and the angular oscillations are quickly die out. For this reason, the pole face bars are called "damper windings or starting windings."

6.4 V-CURVES OF SYNCHRONOUS MOTOR

It has been discussed how variation of field current affects the power factor at which a synchronous motor operates. If now, the load on the motor is held constant while the DC excitation is varied, not only will power factor change, but the input line current also changes.

However, the power factor response of synchronous motor under various DC field excitation currents while holding a constant power can be presented by the V-curves, which are so named because of their distinctive shape. The set of curves is obtained by testing a synchronous motor which is connected to instrumentation and variable mechanical load as shown in Figure 6.3. Data for all meters are recorded at each value of DC excitation while the load is fixed. This allows the power factor of the motor to be determined at each current setting for each load.

When the ac armature current I_a is plotted against the DC field current for no-load, half-load, and full load values respectively, a family of V-curves are obtained. Figure 6.4 shows a typical family of V-curves. In Figure 6.4a, the no-load curve drops

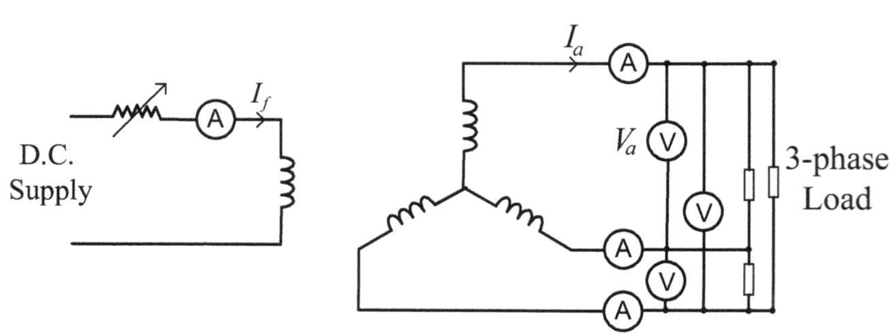

FIGURE 6.3 Circuit diagram for V curves test of synchronous motor.

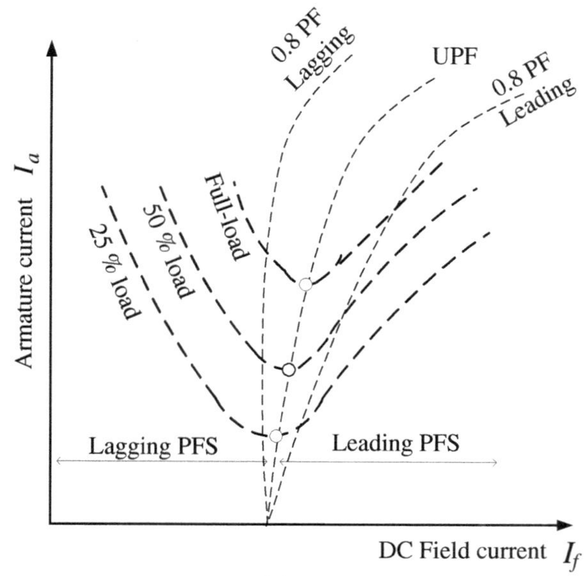

(a)

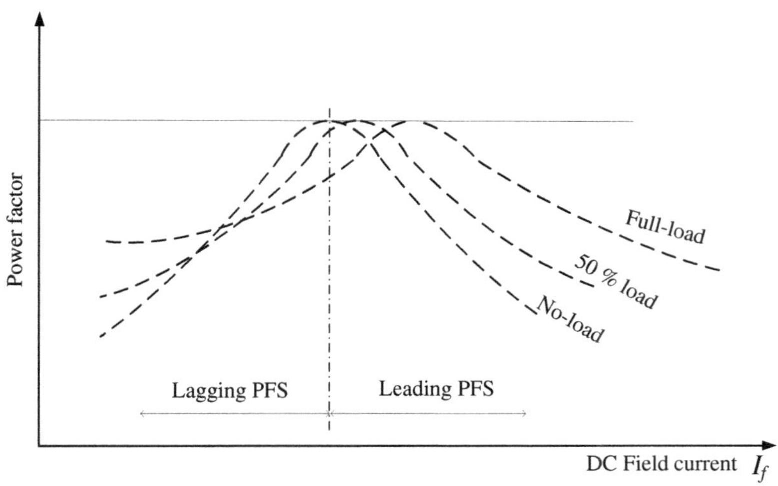

(b)

(a) Armature current vs field current

(b) Power factor vs field current

FIGURE 6.4 Synchronous motor V curves.

to a minimum, but not to zero. The shape of the curves clearly shows that for each load, there is a minimum armature phase current I_a at a specific DC field current I_f.

This specific field current is known as normal excitation. Note that minimum armature current always corresponds to unity power factor. Any increase of field current beyond the normal excitation results in a leading power factor and a higher armature current. Decreasing the field current below that of the normal excitation results in lagging power factor and higher armature current.

6.5 POWER FACTOR CORRECTION

The basic circuit theory shows that for a given electrical load a reduction in power factor results in an increased current. The increased current results in greater copper losses in each component in the circuit. So, it is preferable to bring the load power factor as close to unity as is possible.

6.5.1 Synchronous Condenser

A synchronous motor takes a leading current when over-excited and, therefore, behaves as a capacitor. An over-excited synchronous motor running on no-load in known as synchronous condenser. When such a machine is connected in parallel with induction motors or other devices that operate at low lagging power factor, the leading kVAR supplied by the synchronous condenser partly neutralizes the lagging reactive kVAR of the loads. Consequently, the power factor of the system is improved. Figure 6.5 shows the power factor improvement by synchronous condenser method.

6.6 APPLICATIONS OF SYNCHRONOUS MOTOR

Due to their special operating characteristics, synchronous motor applications usually result in economical and operational advantages to end users. Included in the economic advantages of using synchronous motors are:

- High efficiency
- Power factor correction

In addition to that, there are other specific operational advantages of using synchronous motors as follows:

- Special starting characteristics
- Constant speed under load variation
- Reduced maintenance cost

These motors find application in driving low speed compressors, slow speed fans, pumps, ball mills, metal rolling mills, and process industries.

Example 6.5: A 3-Phase synchronous motor is worked on rated values of 400 V/Ph, 32 A/Ph, and Unity p.f. $X_s = 10\Omega$. Neglecting the value of Ra, calculate E_a and δ.

SOLUTION:

The terminal voltage (V) = OA = 400 V

$$I_a X_S = AB = 32 \times 10 = 320 \text{ V}$$

$$E_a = \sqrt{400^2 + 320^2} = 512.25 \text{ V}$$

The load angle or torque angle $(\cdot) = \tan^{-1}(320/400) = 38.66$ (negative value)

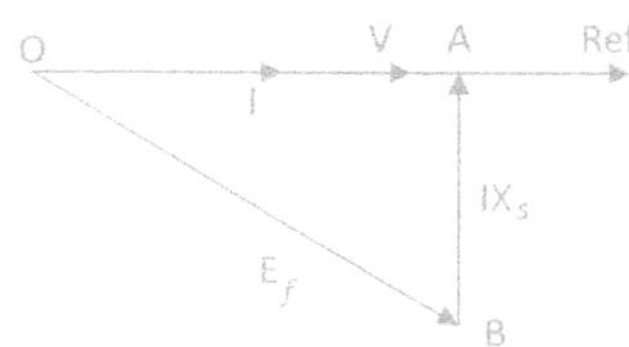

Example 6.6: A 75 kW, 3-phase, Y-connected, 50 Hz, 4-pole, 440 V, cylindrical rotor synchronous motor operates at rated condition with 0.8 p.f. leading. The motor efficiency is 95% and $X_S = 2.5\Omega$. Calculate (i) input power, (ii) armature current, (iii) back EMF, (iv) power angle, (v) electromagnetic power, and (vi) maximum or pull-out torque of the motor.

SOLUTION:

$$P_{out} = 75000 \text{ W}, N_s = 60 \times 50/2 = 1500 \text{ rpm}$$

$$\omega_s = 2\pi \times 1500/60 = 157.08 \text{ rad/s}$$

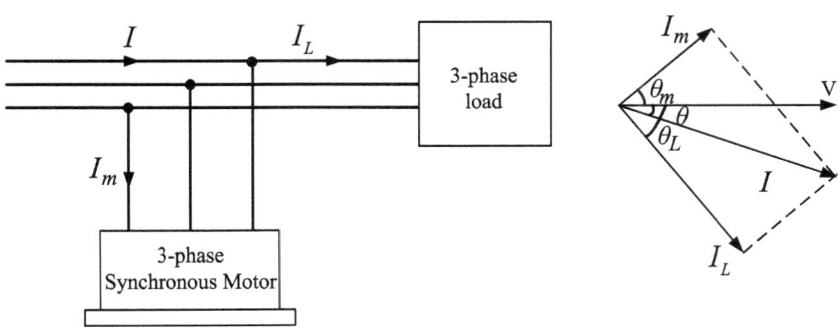

FIGURE 6.5 Synchronous condenser.

(i) the motor efficiency $\eta = P_{out}/P_{in}$

$$0.95 = \frac{75000}{P_{in}}$$

$$P_{in} = 78947.368 \text{ W}$$

ii) $P_{in} = \sqrt{3}\ V\ I_a \cos(\varphi) = 78947.368$ W

$$I_a = \frac{7894.368}{\sqrt{3}(440)(0.8)} = 129.489 \text{ A}$$

(iii) Based on horizontal and vertical analysis for leading power factor motor:

$$\text{Tan}(\delta) = \frac{I_a X_s \cos\phi}{V + I_a X_s \sin\phi}$$

$$\delta = \tan^{-1}(0.5777) = 30^0$$

(vi) From vertical analysis

$$E_a \sin(\delta) = I_a X_s \cos(\varphi)$$

Then

$$E_a = \frac{I_a X_s \cos\phi}{\sin\delta} = 517.956 \text{ V}$$

(v) the electromagnetic power P_{em} is given as

$$P_{em} = 3\ I_a E_a \cos(\delta + \varphi) = 3 \times 517.956 \times 129.489 \cos(30 + 36.87)$$

$$= 79039.1 \text{ W}$$

(vi) the pull out torque occurs at $\delta = 90$

$$T_{pull\_out} = \frac{3V\ E_a \cos\phi}{\omega_s X_s} \sin 90 = 1005.183 \text{ Nm}$$

Example 6.7: A 3-phase, 208 V synchronous motor has a synchronous reactance of 1.0 Ω per phase and negligible armature resistance. The motor draws 50 kVA at 0.8 power factor leading.

Calculate: (a) The stator current I_a and
(b) The excitation voltage E_a.

SOLUTION:

(a) V_t as reference phasor.

$$V_t = \frac{208}{\sqrt{3}} = 120 \text{ V}$$

The stator current is computed as

$$I_a = \frac{50000}{\sqrt{3} \times 208} \angle \cos^{-1} 0.8 = 138.8\angle 36.8 \text{ A}$$

(b) The excitation voltage is found as follows:

$$E_a = V_t - I_a(r_a + jX_s)$$

$$E_a = 120\angle 0 - 138.8\angle 36.8(0{+}j1)$$

$$= 231.7\angle -26.8 \text{ V}$$

Example 6.8: A 3-phase, 200 hp, 480 V, six-pole, 60 Hz, synchronous motor has a synchronous reactance of 1.5Ω/phase. Neglect all losses. The motor is connected to an infinite bus, and it delivers its rated horsepower at unity power factor. Determine the pullout torque, that is, the maximum torque the motor can deliver without losing synchronism.

SOLUTION:

Take the line-to-neutral terminal voltage V_t as reference phasor:

$$V_t = \frac{480}{\sqrt{3}} = 277.1\angle 0.\text{V}$$

At unity power factor, the stator current is in phase with the applied voltages thus:

$$I_a = \frac{200^{*}746}{\sqrt{3} \times 480} \angle 0 = 179.46\angle 0 \text{ A}$$

$$E_a = V_t - I_a(r_a + jX_s)$$

$$E_a = 277.1\angle 0\text{-}179.46\angle 0\ (0 + j1.5)$$

$$= 386.3\angle\text{-}44.2 \text{ V}$$

Since the synchronous motor is assumed to be lossless.

$$P_{max} = T_{max}\omega_m$$

The maximum power occurs at a power angle $\delta = 90^0$

$$P = \frac{3V_t E_a}{X_s}\sin 90 = \frac{3*277.1*386.3}{1.5} = 214 \text{ kW}$$

The mechanical angular speed is found from the radian frequency as

$$\omega_m = (2/p)\omega = (2/6)(2\pi/60) = 125.7 rad/s$$

Finally,

$$T_{max} = P_{max}/\omega_m$$

$$T_{max} = 214{,}000/125.7 = 1702.5 \text{ Nm}$$

Example 6.9: A 200 hp, 2,300 V, 3-phase, 60 Hz, cylindrical-rotor motor has a synchronous reactance of 12 Ω per phase and negligible armature resistance. When it is delivering its rated output, the motor's efficiency is 90% and its power angle δ is 17°.
Determine

(a) The excitation voltage E_a
(b) The stator current I_a and power factor in the power flow diagram for a synchronous motor.

Solution:

(a) The line-to-neutral terminal voltage is taken as reference phasor, thus,

$$V_t = \frac{2300}{\sqrt{3}}\angle 0^0 = 1328\angle 0^0 \text{ V}$$

From the power flow diagram, the input power is the same as the mechanical power developed, since the stator copper loss is negligible. Hence,

$$\eta = P_{out}/P_{in}$$

$$P_{in} = \frac{P_{out}}{\eta} = \frac{3V_t E_a}{X_s}\sin\delta$$

$$= \frac{3*1328*E_a}{12}\sin 17$$

$$= 1708 \text{ V}$$

Therefore, the line-to-line excitation voltage is

$$E_a = 2958\ V(line\text{-}to\text{-}line)$$

(b) Since the armature resistance is negligible, the stator current is given by

$$I_a = \frac{V_t - E_a}{X_s} = \frac{1328\angle 0^0 - 1708\angle 17^0}{j12}$$

$$= 48.8\angle 31.4^0\,\text{A}$$

The power factor is 0.85 leading.

PROBLEMS

6.1 A 3-phase, delta-connected, 220 V, 50 Hz, 1,500 rpm, synchronous motor has a synchronous reactance of 4 Ω/phase. It receives an input line current of 30 A at a leading p.f. of 0.8. Find the line value of induced EMF (E_f) and the load angle expressed in mechanical degrees.
Now, if the mechanical load is disconnected without change of excitation, determine the magnitude of the line current under the new condition. Neglect any losses.

6.2 A 3-phase, Y-connected, 4-pole, 400 V, 200 hp, synchronous motor has a synchronous reactance of 0.5 Ω/phase, the resistance may be neglected. Calculate the load angle and the input current and p.f. when the machine is working at full load with its EMF is adjusted to 1.0 pu based on terminal voltage.

6.3 A 1,000 kVA, 6.6 kV, 50 Hz, 3-phase, 6-pole, Y-connected, synchronous motor has a synchronous impedance of 50 Ω/phase. The excitation is adjusted so that 50% overload is possible before it pulls out of synchronism. Calculate the necessary E_f to permit this overload margin and the output power and the input power factor.
If the output power is held constant and the motor runs at full load, calculate the input power factor and the load angle.

7 Poly-Phase Induction Motors

7.1 INTRODUCTION

Of the many types of motors used today, the one that has the widest industrial application is the poly-phase induction motor. Of all the types of motor studied thus far, the squirrel-cage induction motor (SCIM) is the simplest in construction. This construction produces many advantages including maintenance-free operation, application in remote locations, and operation in hostile environments where dust and other abrasive materials are effective elements. The fact that the other rotor current arises from induction, rather than construction, is the basis for the name of this class of machines. However, they are, sometimes, called "asynchronous" machine because their operational speed is lightly less than synchronous speed. The main advantages of induction motors are:

1. The induction motor is rugged, inexpensive and easy to maintain;
2. Induction motors range in size from a few watts to values on the order of 10,000 hp.;
3. The speed of an induction motor is nearly constant. The speed typically varies only by a few percentage going from no load to rated load.

Some disadvantages of the induction motor are:

1. The speed is not easily controlled;
2. The starting current when motor is directly switched to the supply is relatively high;
3. The power factor is low and lagging.

7.2 CONSTRUCTION OF INDUCTION MOTORS

Like most motors, an AC induction motor has a fixed outer portion, called the stator and a rotor that spins inside with a carefully engineered air gap between the two. Virtually all electrical motors use magnetic field rotation to spin their rotors. A 3-phase AC induction motor is the only type where the rotating magnetic field is created naturally in the stator because of the nature of the supply. DC motors depend either on mechanical or electronic commutation to create rotating magnetic fields. A single-phase AC induction motor depends on extra electrical components to produce this rotating magnetic field. Two sets of electromagnets are formed inside any motor. In an AC induction motor, one set of electromagnets is formed in the stator because

DOI: 10.1201/ 9781003795841-7

of the AC supply connected to the stator windings. The alternating nature of the supply voltage induces an Electromagnetic Force in the rotor (just like the voltage is induced in the transformer secondary) as per Lenz's law, thus generating another set of electromagnets; hence the name – induction motor. Interaction between the magnetic field of these electromagnets generates twisting force, or torque. As a result, the motor rotates in the direction of the resultant torque.

7.2.1 Stator

The stator provides the space for housing for the 3-phase stator windings and also forms the path for the rotating magnetic field. They are built up of thin sheets of thickness (called stampings or laminations) of 0.35 mm to 0.65 mm with of a special core of steel, insulated one from the other by means of paper. The gap facing inner circumference of the plates have suitable slots punched out, either open, semi-closed or completely closed. Normally, the stator core has semi-closed slots where the number of slots. Every part of the stator core is subject to alternate changes in polarity of the magnetic field (due to its rotating nature) so hysteresis losses and eddy-current losses take place. To reduce the former, about 3–5% of silicon is added to high grade steel and to reduce the latter, a large number of thin laminations are stacked together.

7.2.2 Rotor

There are two types of rotors used in induction motors:

1. Squirrel-cage rotor: This rotor consists of a core made of the same thin laminations of silicon steel that the stator uses, but instead of copper windings has several bars which run along its length which are connected to end rings. If the steel laminations were to be removed, these bars look like a squirrel-cage, hence the name, Squirrel-Cage Induction Motor (SCIM). No insulation is required between the core and the bars. This because the very low voltages generated in the rotor bars.
2. Wound rotor: An alternate design is used when variable speed is required. In this case the rotor has the same number of poles as the stator and the windings are made of wire, connected to slip rings on the shaft. Carbon brushes connect the slip rings to the external controller such as variable resistor that allows changes the motor's slip rate.

Compared to squirrel-cage rotors, wound rotor motors are expensive and require maintenance of the slip rings and brushes. The rotor is skewed for these reasons:

1. Helps in reduction of magnetic hum, thus keeping the motor quiet;
2. Helps to avoid "Cogging," that is, locking tendency of the rotor. The tendency of rotor teeth remaining under the stator teeth due to the direct magnetic attraction between the two;
3. Increase in effective ratio of transformation between stator & rotor;

FIGURE 7.1 Types of rotors: (a) squirrel-cage rotor; (b) wound rotor.

4. Increased rotor resistance due to comparatively lengthier rotor conductor bars; and
5. Increased slip for a given torque.

7.3 PRINCIPLES OF OPERATION

It has been proved in Chapter 3 that a rotating and constant resultant magnetic field rotating at synchronous speed is produced by a 3-phase winding displaced in space on an armature if the current flowing through the windings are also displaced in time.

However, the developed rotating magnetic field in the air gap of an induction motor will link the rotor conductors which are at a standstill condition. An EMF will be induced in the rotor windings according to Faraday's law. This induced EMF will act on the short circuit winding of the rotor and result in a current flow in the direction such that the flux developed will oppose the cause which is the rotating air gap flux (according to Lenz's law). This means that there will be two opposite magnetic fields in the air gap of an induction motor, one due to the stator current and which is rotating at synchronous speed while the other is due to the rotor current and is attached to the rotor. Since the rotor is subjected to rotation and magnetic attraction is developed between the opposite poles of the two fields, the rotor will rotate in the same direction of the rotating magnetic field to keep the magnetic coupling acting.

It must be noticed that the rotor of the induction motor cannot rotate at the same speed as the rotating field. The reason is that there must be a relative velocity between the rotating magnetic field and the other rotor conductors in order to induce an EMF in this conductors. The previous analysis is concluded in the following facts: When an induction motor is switched to a 3-phase supply, its rotor will rotate in the same direction of rotation as the air gap magnetic field; and it will rotate at a speed less than the synchronous at which the air gap magnetic field is rotating. Also if two terminals of the stator winding are interchanged, the direction of rotation of the air gap magnetic field will be reversed and consequently the direction of rotor rotation.

7.3.1 A Worked Example

The principle of the induction motor operation is best illustrated by the permanent magnet and aluminum disc shown in Figure 7.2.

In this set up, a permanent magnet is suspended from a string above an aluminum disc which pivots on a bearing set in a fixed iron plate. The field of the permanent magnet is thus completed through the iron plate. As the magnet rotates, the disc below will rotate in the same direction. The disc will rotate in the same direction. This disc will rotate because of the eddy current produced by the relative motion between the disc and the magnetic field crossing it. By Lenz's law, the direction of the eddy currents produces a field tending to oppose the cause which is the permanent magnet field.

So as shown in Figure 7.2b, the induced eddy currents tend to produce a unit S pole in the disc at a point under the rotating N pole, and a unit N pole in the disc under the rotating S pole of the permanent magnet. As long as the permanent magnet continues to rotate, therefore, it will continue to produce eddy currents and poles of opposite polarity in the disc below. A magnetic tie will be developed between the magnet and the disc, and the disc rotates in the same direction as the magnet, but it must rotate at a speed less than that of the magnet. Therefore, the disc speed can never equal to the speed of the magnet. If it did, the induced current would be zero, and no magnetic tie or rotation would be produced. Thus, it must "slip" back in speed in order to produce torque.

7.4 THE CONCEPT OF SLIP

From the previous section, it is realized that the magnitude and frequency of the rotor induced EMF depend on the speed of the relative motion between rotor and the rotating air gap flux. As term called " slip speed," which expresses the speed of the rotor relative to the field, is given by

$$\text{slip speed} = S_s - S_r \text{ rpm} \tag{7.1}$$

where S_s is the speed at which air gap flux rotates

$$S_s = \frac{60f}{p}$$

S_r is speed of rotor in rpm

It is the slip speed difference from the synchronous speed that the induction motor rotor sees. The small speed difference is sufficient to generate a low but real voltage in the rotor conductors. This low voltage, and at the same time low frequency, is sufficient to cause a large circulating currents flowing to the rotor. Although there are few conductors in the rotor circuit, the currents are correspondingly high and motor action is present and adequate. However, the slip speed is usually expressed as a per unit of the synchronous speed. The per unit slip, usually called "the slip" is a

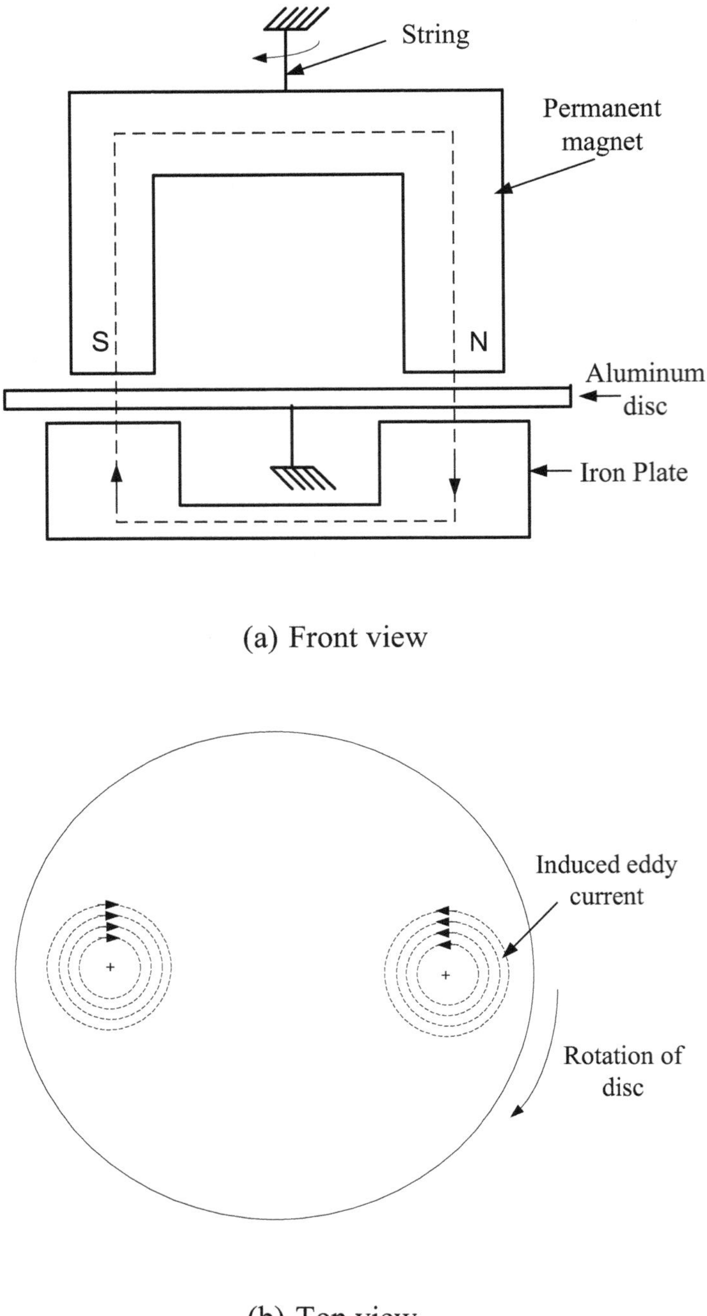

FIGURE 7.2 Induction motor principles.

very useful quantity in studying induction motors. It is represented by symbol s and is defined as follows:

$$s = \frac{S_s - S_r}{S_s} \tag{7.2}$$

Example 7.1: The rotor speed of 4-pole, 50 Hz, induction motor is 1,350 rpm. Calculate the per unit slip.

SOLUTION:

The synchronous speed

$$S_s = \frac{60 \times 50}{2} = 1500\ r/\min$$

The slip

$$s = \frac{1500 - 1350}{1500}$$

Example 7. 2: A 220 V, 3-phase, 6-pole, 50 Hz, induction motor is running at a slip of 3.5%. Find:

(a) The speed of the magnetic fields in revolutions per minute
(b) The speed of the rotor in revolutions per minute
(c) The slip speed of the rotor
(d) The rotor frequency in hertz

SOLUTION:

(a) The speed of the magnetic fields is

$$N_s = 120 \times \frac{50}{6} = 1000\ \text{rpm}$$

(b) The speed of the rotor is

$$N_m = (1 - s)N_s = 965\ \text{rpm}$$

(c) The slip speed of the rotor is

$$N_{slip} = sN_s = 0.035 \times 1000 = 35\ \text{rpm}$$

(d) the rotor frequency

$$f_r = \frac{N_{slip} P}{120} = 1.75\ \text{Hz}$$

7.5 THE FREQUENCY OF ROTOR VOLTAGES AND CURRENT

The frequency of supply to the stator is f Hz. The EMF induced in the rotor when the rotor is rotating at a speed S_r is due to the difference of speed of the rotating field and the rotor speed, that is, S_s-S_r.

When the rotor is at standstill, that is, not yet rotating, the difference between speed of rotation of the stator field and the rotor is $S_s - 0 = S_s$. The EMF induced in the rotor will be maximum and will have a frequency f. When the rotor attains a speed of S_r, the frequency of the induced EMF will get reduced, and let this rotor frequency be called, f_r f_r. Thus, when the relative speed is S_s, the frequency of the induced EMF in the rotor is f. When the relative speed is $(S_s$-$S_r)$, the frequency is f_r. The relationship between f and f_r is found as follows:

$$f_r = sf_s \tag{7.3}$$

7.5.1 Rotor Reactance

The rotor circuit is made of heavy bars of short length which are short circuited at both ends. So, the rotor circuit has relatively low resistance. However, since the bars are embedded in iron, they possess the property of inductance and, consequently, inductive reactance. The magnetic circuit of the rotor is the same at standstill and when rotating, since none of the physical quantities involved are changed. The inductance of the rotor, L_r, is therefore constant. Rotor frequency, however, does vary, and it is obvious that the inductive reactance of the rotor varies directly with rotor frequency, that is, $X_r = 2\pi f_r L_r$. The rotor frequency is directly proportional to slip (7.3), and the rotor reactance therefore varies in the same manners. Thus

$$X_r = sX_{lr} \tag{7.4}$$

Thus at any slip s,

$$\text{Rotor reactance/phase } = \text{s}X_{lr}$$

$$\text{Rotor EMF/phase} = \text{s}E_2$$

where X_{lr} and E_2 are the locked rotor reactance and rotor EMF at standstill.

7.5.2 The Rotor Current

Figure 7.3 shows the circuit of a 3-phase induction motor at any slip s. The rotor is assumed to be of wound type and star connected. Note that rotor EMF/phase and rotor reactance/phase are sE_2 and sX_{lr} respectively. The rotor resistance/phase is R_r and is independent of frequency and, therefore, does not depend upon slip. Likewise, stator winding values R_s and X_s do not depend upon slip.

Since the motor represents a balanced 3-phase load, we need consider one phase only; the conditions in the other two phases being similar.

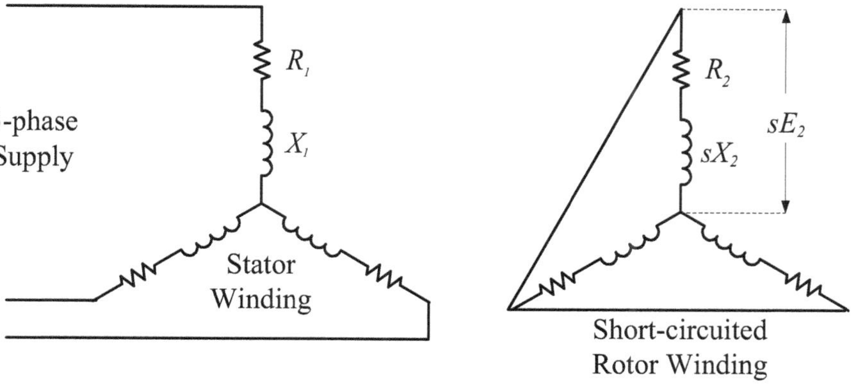

FIGURE 7.3 The circuit of induction motor.

$$\text{Rotor current/phase, } I_r = \frac{E_r}{Z_r} = \frac{E_r}{\sqrt{R_r^2 + X_r^2}} \tag{7.5}$$

$$\text{Rotor p f., } \cos\phi_r = \frac{R_r}{Z_r} = \frac{R_r}{\sqrt{R_r^2 + X_r^2}} \tag{7.6}$$

When running at slip s, Figure 7.4 shows one phase of the rotor circuit when the motor is running at slip s.

$$\text{Rotor current/phase, } I_r = \frac{sE_r}{Z_r} = \frac{sE_r}{\sqrt{R_r^2 + (sX_r)^2}} \tag{7.7}$$

$$\text{Rotor p f., } \cos\phi_r = \frac{R_r}{Z_r} = \frac{R_r}{\sqrt{R_r^2 + (sX_r)^2}} \tag{7.8}$$

7.6 THE EQUIVALENT CIRCUIT OF AN INDUCTION MOTOR

An induction motor relies for its operation on the induction of voltages and currents in its rotor circuit from the stator circuit (transformer action). This induction is essentially a transformer operation; hence the equivalent circuit of an induction motor is similar to the equivalent circuit of a transformer.

By assuming that the MMF of the load current I_2 balances the MMF of the rotor current, the combined equivalent circuit of an induction motor is presents in Figure 7.5.

As in any transformer, there is certain resistance and self-inductance in the primary (stator) windings, which must be represented in the equivalent circuit of the machine. They are – R_1 – stator resistance and X_1 – stator leakage reactance Also, like any transformer with an iron core, the flux in the machine is related to the integral of the applied voltage E_1.

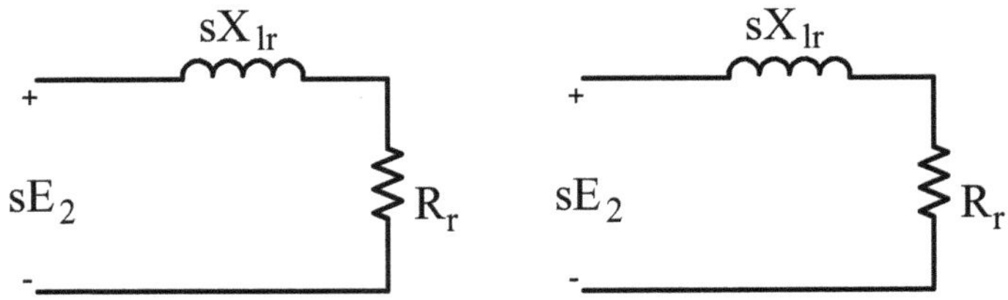

FIGURE 7.4 Rotor equivalent circuit.

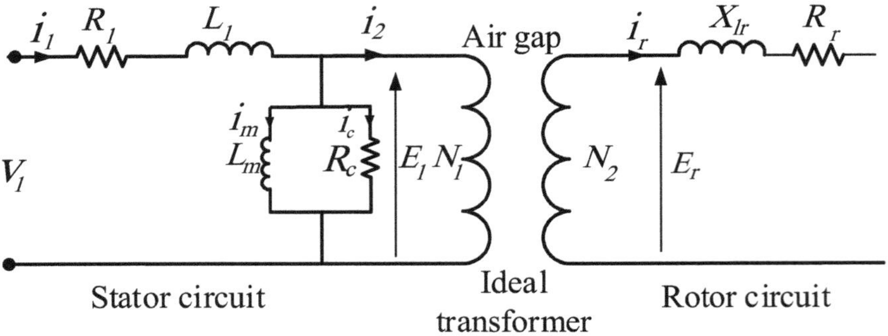

FIGURE 7.5 The transformer model of an induction motor.

The voltage E_1 is the induced EMF in the stator and is given by:

$$E_1 = 4.44 fs\phi T_{ph} K_w \tag{7.9}$$

It should be noted that the stator input current I_1 can be resolved into two components, the load current I_2 and the no-load current I_0.

The primary internal stator voltage E_1 is coupled to the secondary E_r by an ideal transformer with an effective turn ratio "a." The turns ratio for a wound rotor is basically the ratio of the conductors per phase on the stator to the conductors per phase on the rotor. It is rather difficult to see "a" clearly in the cage rotor because there are no distinct windings on the cage rotor. E_r in the rotor produces current flow in the shorted rotor (or secondary) circuit of the machine. The primary impedances and the magnetization current of the induction motor are very similar to the corresponding components in a transformer equivalent circuit.

In Figure 7.5, two different electrical circuits are available which can be presented by the following equations:

$$V_1 = (r_1 + jx_1)I_1 + E_1 \tag{7.10}$$

$$E_r = (\frac{R_r}{s} + jX_{1lr})I_r \tag{7.11}$$

To produce the final per-phase equivalent circuit for an induction motor, it is necessary to refer the rotor part of the model over to the stator side. In an ordinary transformer, the voltages, currents, and impedances on the secondary side can be referred to the primary by means of the turns ratio of the transformer.

Exactly the same sort of transformation can be done for the induction motor's rotor circuit. If the effective turns ratio of an induction motor is "a," then the transformed rotor current becomes

$$I_2' = \frac{I_r}{a} \tag{7.12}$$

And the rotor impedance:

$$Z_2' = a^2\left(\frac{R_r}{s} + jX_r\right) \tag{7.13}$$

If we make the following definition

$$R_2' = a^2 R_r \tag{7.14}$$

$$X_2' = a^2 X_r \tag{7.15}$$

The final per-phase equivalent circuit is as shown in Figure 7.6.

7.7 DETERMINATION OF EQUIVALENT CIRCUIT PARAMETERS

It is possible to find the parameters of the equivalent circuit of the 3-phase induction motor experimentally as shown in Figure 7.6. For this purpose, three tests should be conducted:

1- **The DC Test:**
The DC resistance of the stator can be measured by applying DC current to the terminals of the winding of each phase and taking the reading of the

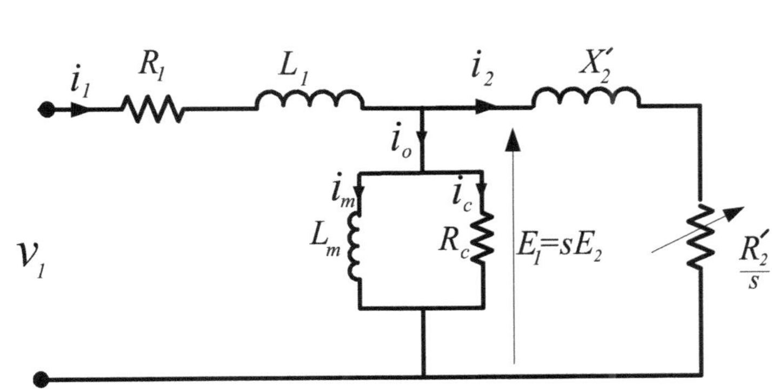

FIGURE 7.6 The per-phase equivalent circuit of an induction motor.

voltage and the current (or using ohmmeter) and determine the DC resistance as follows:

$$R_{DC} = \frac{V_{DC}}{I_{DC}} \tag{7.16}$$

Then, the AC resistance is given by:

$$R_1 = 1.5R_{DC} \tag{7.17}$$

2- **The Locked Rotor Test:**

When the rotor is locked (i.e., prevented from running), s is equal to 1. The secondary impedance becomes much less than the magnetizing branch and the corresponding equivalent circuit becomes that of Figure 7.7. The readings to be obtained from this test are:

a) 3-phase power $P_{3\phi BL}$

b) Line voltage V_{L_BL}

c) Line current I_{BL}

From these readings, the per phase values of the power P_{BL} and phase voltage V_{BL} can be obtained as follows:

$$P_{BL} = \frac{P_{3\phi\_BL}}{3} \tag{7.18}$$

$$V_{BL} = \frac{V_{L\_BL}}{\sqrt{3}} \tag{7.19}$$

Then, R_{eq}, Z_{eq}, and X_{eq} can be obtained using the following equations:

$$R_{eq} = \frac{P_{BL}}{I_{BL}^2} \tag{7.20}$$

$$Z_{eq} = \frac{V_{BL}}{I_{BL}} \tag{7.21}$$

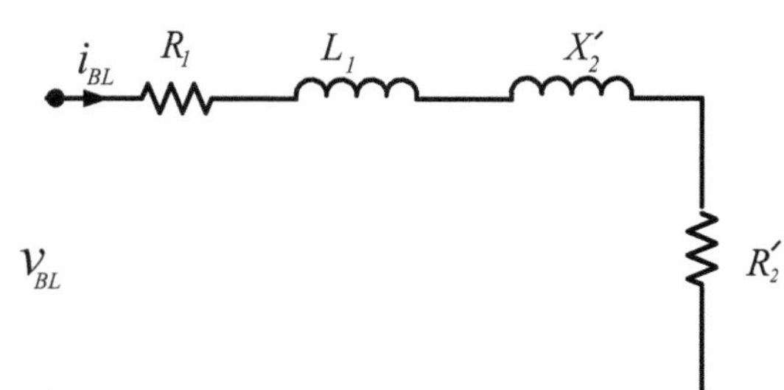

FIGURE 7.7 Approximate equivalent circuit for the locked rotor conditions.

$$X_{eq} = \sqrt{Z_{eq}^2 - R_{eq}^2} \tag{7.22}$$

Separation of X_1, X_2', R_1 and R_2' can be done as follows:

$$X_1 = X_2' = \frac{1}{2} X_{eq} \tag{7.23}$$

$$R_2' = R_{eq} - R_1 \tag{7.24}$$

3- **The No Load Test**

When the induction motor runs at no load, the rotor speed approaches the synchronous speed. The slip becomes very small in this case. Accordingly, the secondary impedance becomes high compared with the magnetizing branch. the equivalent circuit can be approximated by that of Figure 7.8. The readings to be obtained from this test are:

a) 3-phase power $P_{3\phi NL}$
b) Line voltage V_{L_NL}
c) Line current I_{NL}

From these readings, the per phase values of the power P_{NL} and phase voltage V_{NL} can be obtained as follows:

$$P_{NL} = \frac{P_{3\phi\_NL}}{3} \tag{7.25}$$

$$V_{NL} = \frac{V_{L\_NL}}{\sqrt{3}} \tag{7.26}$$

Then, R_{e+h} and X_m can be obtained using the following equations:

$$P_{core+mechanical} = P_{NL} - I_{NL}^2 R_1 \tag{7.27}$$

$$E_1 = V_{NL} - I_{NL}(R_1 + jX_1) \tag{7.28}$$

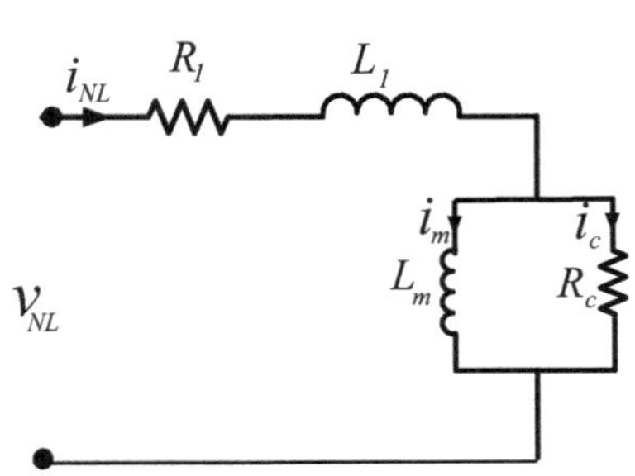

FIGURE 7.8 Approximate equivalent circuit for the no load conditions.

$$R_c = \frac{|E_1^2|}{P_{core + mechanical}} \tag{7.29}$$

$$I_c = \frac{E_1}{R_c} \tag{7.30}$$

$$I_m = \sqrt{I_{NL}^2 - I_c^2} \tag{7.31}$$

$$X_m = \frac{E_1}{I_m} \tag{7.32}$$

7.8 INDUCTION MOTOR PERFORMANCE

The induction motor performance can be obtained from two approaches; the first depends on the equivalent circuit of the motor and the other approach depends on the power flow principles inside the motor. These two approaches are discussed in detail in the following sections:

7.8.1 The Equivalent Circuit Approach

While the equivalent circuit affords a convenient basis for the performance calculations when the numerical values of the motor constants are known, it leads to general derivations in the different in which the effect of variations in the different parameters of the motor characteristics can be computed and studied. The exact equivalent circuit shown in Figure 7.6 will be used in the following analysis:

$$V_1 = I_1 z_1 + E_1 \tag{7.33}$$

and since

$$I_1 = I_2 + I_0 \tag{7.34}$$

Equation (7.33) can be rewritten in the form

$$\begin{aligned} V_1 &= E_1 + (I_2 + I_0) z_1 = E_1 + I_2 z_1 + I_0 z_1 \\ &= I_2 z_2 + I_2 z_1 + I_0 z_1 = I_2 (z_2 + z_1) + I_0 z_1 \end{aligned}$$

$$I_0 = \frac{E_1}{z_m}$$

Thus,

$$V_1 = (z_1 + z_1)\, I_2 + \frac{z_1}{z_m} E_1 \tag{7.35}$$

Since the reactive component in z_1 is several times greater than the resistive component in the conventional induction motor and E_1 is approximately equal to the in-phase component of V_1

$$\frac{z_1}{z_m}E_1 \cong \frac{x_1}{x_m}E_1 \tag{7.36}$$

By substituting Equation (7.36) into Equation (7.35) we get:

$$V_1\left(1-\frac{x_1}{x_m}\right)=\left[\left(r_1+\frac{r_2}{s}\right)+j(x_1+x_2)\right]I_2 \tag{7.37}$$

Equation (7.37) will be used to calculate the load component of the stator current I_2 which is used to obtain the motor performance. From the exact equivalent circuit, the real part of the rotor power input:

$$P_{gap}=3I_2^2\frac{r_2}{s} \tag{7.38}$$

The power consumed in the rotor resistance:

$$P_{cu\_rotor}=3I_2^2r_2 \tag{7.38}$$

The developed mechanical power must be the difference between the power input and the power that transferred to heat in the rotor winding. Accordingly, the mechanical power developed

$$P_d=3I_2^2r_2\left(\frac{1}{s}-1\right)=3I_2^2r_2\left(\frac{1-s}{s}\right) \tag{7.40}$$

The torque developed is the developed power divided by the mechanical angular speed of the rotor

$$T_d=\frac{P_d}{\omega_r}\ \text{Nm} \tag{7.41}$$

where

$$\omega_r=\frac{2\pi S_r}{60}\ \text{rad/sec} \tag{7.42}$$

Example 7.3: A 150 hp, 440V, 3-phase, 50 Hz, 8-pole, wound-rotor induction motor has its stator and rotor both connected in a Y configuration. The ratio of effective stator turns to effective rotor turns a= 2.4:1. The motor parameters per phase are:

$$R_l=0.52\Omega,\ X_l=1.15\Omega,\ R_r=0.11\Omega,\ X_r=0.2\Omega,\ X_m=40\Omega.$$

The windage and friction losses (assumed constant) $P_{cu\_rotor} = 220W$

The stray load losses $P_{stray} = 120W$

For a slip s= 0.045, and by using the motor equivalent circuit calculate: (a) stator current, (b) motor power factor, (c) current in the rotor winding, (d) output horse power, (e) efficiency, and (f) developed torque.

Solution: The motor current referred to the stator

$$I_2 = \frac{V(1 - x_1/x_m)}{z_1 + z_2}$$

$$r_2 = a^2 R_r = (2.4)^2 \times 0.11 = 0.634\Omega/phae \;\; I_2$$

$$x_2 = a^2 X_r = (2.4)^2 \times 0.2 = 1.15\Omega/phae$$

$$z_2 = \frac{r_2}{s} + jx_2 = \frac{0.634}{0.045} + j1.15 = 14.1 + j1.15$$
$$= 14.13\angle 4.7^0\,\Omega/phase$$

$$z_1 = r_1 + jx_1 = 0.52 + j1.15 = 1.26\angle 65.6^0\,\Omega/phase$$

$$I_2 = \frac{\frac{440}{\sqrt{3}}\left(1 - \frac{1.15}{40}\right)}{14.62 + j2.3} = \frac{246.7}{14.8\angle 9^0} = 16.7\angle -9^0\,A$$

The no-load current

$$I_2 = \frac{V_1 - V_2(z_1 + z_2)}{z_1} = 0.69 - j6.17A$$

(a) The stator current is given by

$$I_1 = I_2 + I_0 = 19.3\angle 27^0$$

(b) The power factor of motor = cos 27^0 = 0.89 lag.

(c) The current in the rotor winding $I_r = a\,I_2 = 40.1\,A$

(d) The developed mechanical power (using Equation 7.40)

$$P_d = 3I_2^2 r^2\left(\frac{1}{s} - 1\right) = 3(16.7)^2 \times 0.634 \times \left(\frac{1 - 0.045}{0.045}\right) = 11260W$$

The output power = $P_d - P_{cu\_rotor} - P_{stray}$ = 11260 – $(220 + 120)$
= 10920 W=14.65 hp

(e) Efficiency $= \dfrac{Output}{input}$

Input power to motor = $3\ V_sI_1\cos\theta_m = 3(254)\ (19.3)\ (0.89) = 13080$ W

$$\text{Efficiency} = \frac{10920}{13080} \times 100 = 83.5\%$$

(f) Developed torque $= \dfrac{developed\ power}{\omega_r}$

$$= \frac{11260}{2\pi(1-s)S_r/60} = 83.5\%$$

$$= \frac{11260}{90} = 125.1\ \text{Nm}$$

7.8.2 Power Flow in Induction Motors

This approach is based on the law of conservation of energy which states " in an induction motor the input power is equal to the output power plus sum of losses inside the motor."

The input power is given by

$$P_{in} = 3V_sI_1\cos\theta_m \tag{7.43}$$

Now by following the path of power flow inside the motor, we get:

$$\text{The stator copper losses SCL} = 3I_1^2r_1 3I_1^2r_1$$

Neglecting the stator core losses at this point, the power available at the air gap ready to cross the rotor circuit is called air gap power P_g and is given by:

$$P_g = P_{in} - SCL = 3V_sI_1\cos\theta_m - 3I_1^2r_1 \tag{7.44}$$

The rotor power input which is the air gap power is partly consumed as rotor copper loss (RCL) and the remainder will be the available in the rotor to develop mechanical power. Thus, the developed mechanical power

$$P_d = P_g - RCL = P_g - 3I_2^2r_2 \tag{7.45}$$

The net mechanical output power

$$P_{out} = P_d - P_{rot} \tag{7.46}$$

where,

$$P_{rot} \text{ is the rotational losses} = \text{friction and windage losses } P_{fw} + \text{Stray load loss } P_{LL} + \text{ Core loss}_{Pe+h}$$

From the equivalent circuit shown in Figure 7.6, the air gap power which is the power input to the rotor circuit is given by:

$$P_g = 3I_2^2 \frac{r_2}{s} \tag{7.47}$$

But the rotor copper loss is given by $RCL = 3I_2^2 r_2$

$$\text{Thus RCL} = sP_g \tag{7.48}$$

From equation (7.44), the developed mechanical power is

$$P_d = P_g - RCL = 3I_2^2 \frac{r_2}{s} - 3I_2^2 r_2$$

$$= 3I_2^2 \frac{r_2}{s}\left(\frac{1-s}{s}\right) = 3I_2^2 \frac{r_2}{s}(1-s)$$

$$= P_g(1-s) \tag{7.49}$$

Equations (7.46), (7.47), and (7.48) will lead to the following important relation

$$P_g = P_d : RCL = 1 : 1-s : s \tag{7.50}$$

This useful relation indicates once air gap power is determined, the other two quantities can be obtained at any given slip.

The torque developed by the electromagnetic energy conversion process is given by the developed mechanical power divided by the angular speed of rotor.

$$T_d = \frac{P_d}{\omega_r} \tag{7.51}$$

But since slip

$$s = \frac{S_s - S_r}{S_s} = \frac{\omega_s - \omega_r}{\omega_s}$$

$$\omega_s - \omega_r(1-s) \tag{7.52}$$

Substituting equations (7.48) and (7.51) into Equation (7.50), we get

$$T_d = \frac{P_d(1-s)}{\omega_r(1-s)} = \frac{P_g}{\omega_s} \tag{7.53}$$

The actual output torque is given by

$$T_{out} = \frac{P_{out}}{\omega_r}\ \text{Nm} \tag{7.54}$$

Example 7.4: A 3-phase, 4-pole induction motor draws 30A from a 420V, 50 Hz supply at power factor of 0.82 lagging. The stator copper loss is 1,200 W and the rotor copper loss is 850 W. The friction and windage loss = 1,200 W. The core loss = 800 W. The stray load loss=200 W. Calculate: (a) the air gap power, (b) the developed power, (c) the output horsepower, (d) the efficiency, (e) the slip, (f) the operating speed, (g) the developed torque, and (h) the output torque.

Solution:

(a) $P_g = P_{in} - SCL = \sqrt{3}(420)(30)(0.82) - 1200 = 16695.5\ \text{W}$

(b) $P_d = P_g = P_{in} - SCL = \sqrt{3}(420)(30)(0.82) - 1200 = 16695.5\ \text{W}$

(c) $P_{out} = P_d - P_{rot} = 15845.5 - (250 + 800 + 200)$

$$\text{W}=14595.5$$
$$= 19.5\ \text{hp}$$

(d) $\text{Efficiency} = \frac{P_{out}}{P_{in}} = \frac{14595.5}{17895.5} \times 100 = 81.5\%$

(e) $P_g : RCL = 1 : s$

$$s = \frac{RCL}{P_g} = \frac{850}{16695.5} = 0.05$$

(f) $S_s = \frac{60f}{P} = \frac{60 \times 50}{2} = 1500\ \text{r/min}$

$$S_r = Ss(1-s) = 1500(1-0.05) = 1425\ \text{r/min}$$

(g) $T_d == \frac{P_d}{\omega_r} = \frac{15845.5}{2\pi \frac{S_r}{60}} = 106\ \text{Nm}$

(h) $T_{out} == \frac{P_{out}}{\omega_r} = \frac{14595.5}{2 \times \frac{1425}{60}} = 97.8\ \text{Nm}$

7.9 TORQUE-SPEED CHARACTERISTICS OF INDUCTION MOTOR

The torque developed at standstill is directly proportional to rotor current and strength of the stator flux:

$$T_{st} = K_t I_r \phi \cos\theta_r \tag{7.55}$$

where I_r = rotor current at standstill
ϕ = Instantaneous value of flux
$\cos\theta_r$ = rotor p.f. at standstill
K_t = constant of proportionality

It is now possible to derive an equation for the torque developed at starting (standstill condition) in terms of the rotor parameters.

At standstill, the locked rotor impedance Z_{lr} is

$$Z_{lr} = R_r + jX_{lr} \tag{7.56}$$

and

$$\cos\theta_r = \frac{R_r}{Z_{lr}} = \frac{R_r}{\sqrt{R_r^2 + X_{lr}^2}} \tag{7.57}$$

The current in the rotor at standstill is

$$|I_r| = \frac{E_{lr}}{|Z_{lr}|} = \frac{E_{lr}}{\sqrt{R_r^2 + X_{lr}^2}} \tag{7.58}$$

Now, by substituting Equations (7.56) and (7.57) into Equation (7.54), we get:

$$T_{st} = K_t \phi \frac{E_{lr}}{\sqrt{R_r^2 + X_{lr}^2}} = \frac{R_r}{\sqrt{R_r^2 + X_{lr}^2}}$$

$$= \frac{K_t \phi E_{lr} R_r}{R_r^2 + X_{lr}^2} \tag{7.59}$$

At starting. The rotor is locked, and the motor presents the static transformer case. Because of transformer action, E_{lr} is proportional to ϕ which, in turn, is proportional to the supply voltage assuming a linear magnetization curve. Then Equation (7.58) may be simplified further to

$$T_{st} = \frac{K_t^* V_p^2 R_r}{R_r^2 + X_{lr}^2} \tag{7.60}$$

where K_t^* is a new constant of proportionality.

For given constant rotor parameters and at a constant supply frequency, the starting torque is solely a function of the square of the voltage applied to stator winding.

7.9.1 Developed Torque at any Slip

Once the motor develops a starting torque T_{st} sufficient to accelerate the rotor and to overcome mechanical friction and load, its rotor frequency, rotor reactance, and rotor induced EMF are presented by sf, $s\ X_{lr}$, and $s\ E_{lr}$ respectively. As the rotor frequency and reactance decreases, the rotor induced EMF also decreases as the motor speed rises. For any given slip s, the rotor current is given by:

$$I_r = \frac{E_{lr}}{\sqrt{\left(\frac{R_r}{s}\right)^2 + X_{lr}^2}} \tag{7.61}$$

And the power factor will be

$$\cos\theta_r = \frac{R_r/s}{\sqrt{\left(\frac{R_r}{s}\right)^2 + X_{lr}^2}} \tag{7.62}$$

By substituting Equations (7.60) and (7.61) into Equation (7.54) the torque at any slip is given by

$$T = \frac{K_t\ \phi\ E_{lr} R_r/s}{\left(\frac{R_r}{s}\right)^2 + X_{lr}^2} \tag{7.63}$$

But since the standstill, locked rotor EMF, E_{lr}, is directly proportional to ϕ, the torque may be expressed by

$$T = \frac{K_t^{'}\phi^2 R_r/s}{\left(\frac{R_r}{s}\right)^2 + X_{lr_t}^2} = K_t^{*}V_p^2\frac{R_r/s}{\left(\frac{R_r}{s}\right)^2 + X_{lr}^2} \tag{7.64}$$

7.9.2 Maximum Torque

There are three main values of induction motor torque of specific importance: the starting torque, the maximum torque, and the load torque. The determination of these torque values enables us to fit the motor the most suitable application. The highest torque that a motor (of constant parameters) can develop is known as the maximum torque, the breakdown torque, or the pull-out torque. At this value of torque, the motor has reached a point where it cannot continue to increase its torque. A further increase in load will cause the motor to stall.

To obtain the maximum torque, differentiate sides of Equation (7.63) with respect to s and equate the result to zero, as follows:

$$\frac{dT}{ds} = K_t^{'}\phi^2 \frac{\left(R_r^2 + S^2X_{lr}^2\right)R_r - sR_r\left(2sX_{lr}^2\right)}{\left[R_r^2 + s^2X_{lr}^2\right]^2} = 0 \tag{7.65}$$

From which $R_r^2 + s^2X_{lr}^2 = 0$. Thus

S_m (slip corresponding to maximum torque) $= \dfrac{R_r}{X_{lr}}$

Substitute in Equation (7.63), the maximum torque is found to be

$$T_{\max} == K_t^{*}V_p^2 \frac{S_m R_r}{2\left(S_m X_{lr}\right)} \tag{7.66}$$

The general shape of the torque-slip characteristics for induction motor is shown in Figure 7.9.

The machine will operate as a motor when $1 \geq s \geq 0$, as a generator when $s < 0$ and as brake when $s > 1$. The machine must be driven above synchronous speed for generation operation and must be driven against the direction of rotation of its MMF for operation as a brake.

It must be noted that the slip S_m at which maximum torque is available is a direct function of the motor of the rotor resistance. But the value of maximum torque, given by (7.64), is independent of the rotor resistance. Now, by assuming that the voltage applied to the stator is constant while the rotor resistance is varied, the maximum torque will have the same value for all values of rotor resistance. Both the starting torque and the slip at maximum torque are increased as the rotor resistance is increased as shown in Figure 7.10.

A useful relation that is sometimes employs in the solution of induction motor torque problems is an equation which relates the torque at any speed to maximum or breakdown torque.

$$T = T_{\max} \frac{2}{\dfrac{S_m}{S} + \dfrac{S}{S_m}} \tag{7.67}$$

where T is the torque at any slip.

Example 7.5: Six-pole, 25 hp, 50 Hz, 420 V induction motor has a starting torque of 200 Nm and full load torque of 150 Nm. Calculate

(a) The starting torque when the stator voltage is 300V.
(b) The voltage which must be applied to the stator in order to develop a starting torque equal to the full torque.
(c) The starting current of the motor when the voltage is reduced to 300V if the starting current when rated voltage is impressed on the stator.

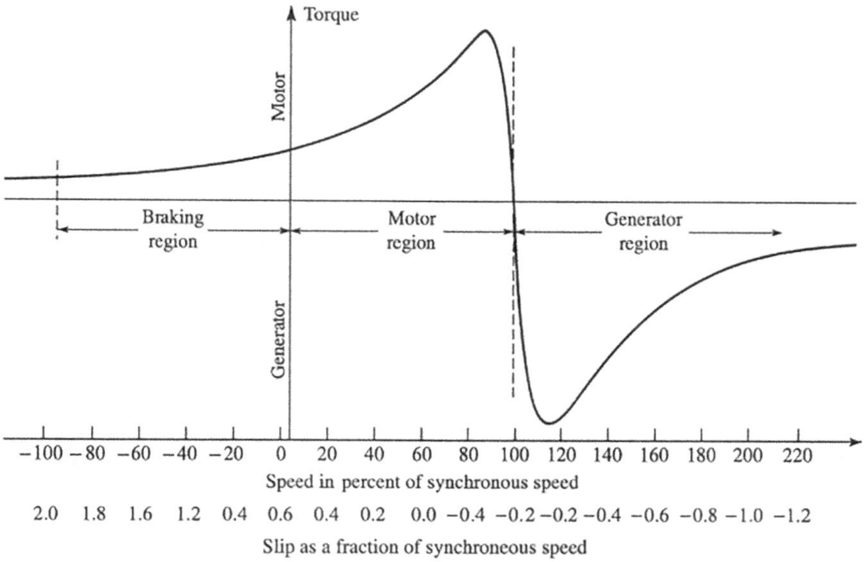

FIGURE 7.9 Torque slip curve.

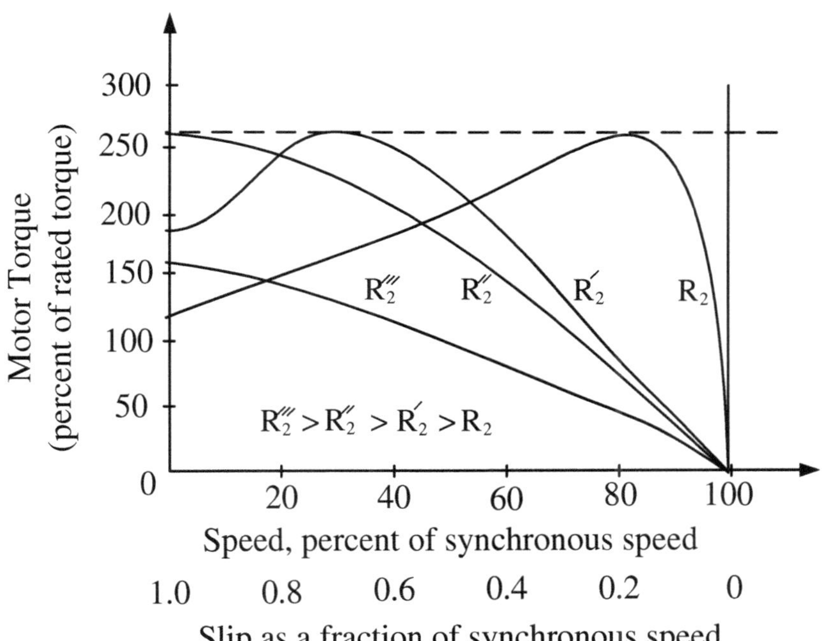

FIGURE 7.10 Effect of changing of rotor resistance on starting and running characteristics of induction motor.

Solution:

$$T = KV^2$$

(a) $(T_{st})_{300V} = (T_{st})_{420V} \cdot (voltagertio)^2$

$$= 200 \times \left(\frac{300}{420}\right)^2 =$$

$$= 102 \text{ Nm}$$

(b) $V_{req \cdot line} = (Torqueratio)^{\frac{1}{2}} \times V_{rated} = \left(\frac{150}{200}\right)^{\frac{1}{2}} \times 420 = 364 \text{ V}$

(c) $I_{1st} = (voltageration) \times I_{1st\ rated}$

$$= \left(\frac{300}{420}\right) \times 120 = 85.7 \text{ A}$$

Example 7.6: A 12-pole, 50 Hz induction motor has full-load speed of 450 r/min. the standstill rotor reactance and resistance per phase are 2 and 6 Ω respectively. Calculate

(a) Slip and speed at which maximum torque is developed
(b) The added rotor resistance to develop maximum torque at starting
(c) The new full-load slip which added to rotor resistance
(d) Speed regulation with added rotor resistance

Solution:

$$S_s = \frac{60f}{P} = \frac{30 \times 50}{6} = 500 \text{ r/min}$$

(a) $S_m = \frac{R_r}{X_{lr}} = \frac{0.6}{2} = 0.3$

$$S_{rat\ \max\ torque} = Ss(1 - S_m) = 500(1 - 0.3) = 350 \text{ r/min}$$

(b) At starting, slip must be unity

$$S_m = 1 \quad \therefore R_r + R_{add} = X_{lr}$$

The external added resistance $R_{add} = 2 - 0.6 = 1.4\Omega$

(c) Original full load slip

$$S_{FL} = \frac{S_s - S_r}{S_s} = \frac{500-400}{500} 0.01$$

New slip (with added rotor resistance) $= S_{FL} \times \frac{(R_r)_{new}}{(R_r)_{origin}}$

$$= 0.1 \times \frac{2}{0.6} = 0.333$$

Example 7.7: The total power supplied to a 3-phase induction motor is 4,200 W and the corresponding stator losses are 350 W. Calculate

(a) The rotor power loss when the slip is 2.5%
(b) Total mechanical power developed
(c) Output horsepower if the friction and windage losses are 120 W
(d) Motor efficiency

SOLUTION:

(a) Input power to rotor P_g = Total input power – stator power loss

$$= 4200 - 350 = 3850 \text{ W}$$

$P_g : RCL = 1 : s$

Rotor copper loss = $s\,P_g = 0.025 \times 3850 = 96$ W

(b) $P_g : P_m = 1 : 1 - s$

Mechanical power developed $P_d := (1\text{-s})P_g$

$$= (1 - 0.025) \times 3850 = 3754 \text{ W}$$

(c) $P_{out} = P_d = P_{fw} = 3754 \text{ -120} = 3634$ W

$$= 4.87 \text{ hp}$$

(d) Efficiency $= \frac{P_{out}}{P_{in}} = \frac{3634}{4200} \times 100 = 86.5\%$

Example 7.8: Find the percentage tapping required on an auto-transformer required for a squirrel-cage motor to start the motor against 1/4 of full-load torque. The short-circuit current on normal voltage is four times the full-load current and the full-load slip is 3%.

SOLUTION:

$$\frac{T_{st}}{T_f} = \frac{1}{4}, \ \frac{I_{sc}}{I_f} = 4, \ S_f = 0.03$$

$$\therefore \text{ Using } \frac{T_{st}}{T_f} = K^2 \left(\frac{I_{sc}}{I_f} \right)^2 S_f, \text{ we get } \frac{1}{4} = K^2 \times 4^2 \times 0.03,$$

$$\therefore \therefore K^2 = \frac{1}{64 \times 0.03} \qquad \therefore \therefore K = 0.722 \quad \text{or}$$

$$K = 72.2\%$$

Example 7.9: A 20 hp (14.92 kW), 400 V, 950 rpm, 3-phase, 50 Hz, 6-pole, cage motor with 400 V applied takes 6 times full-load current at standstill and develops 1.8 times full-load running torque. The full-load current is 30 A.

(a) What voltage must be applied to produce full-load torque at starting?
(b) What current will this voltage produce?
(c) If the voltage is obtained by an auto-transformer, what will be the line current?
(d) If starting current is limited to full-load current by an auto-transformer, what will be the starting torque as a percentage of full-load torque?

Ignore the magnetizing current and stator impedance drops.

SOLUTION:

(a) Remembering that $T \propto V^2$, we have
In the first case, $1.8T_f \propto 400^2$; In the second case, $T_f \propto V^2$

$$\left(\frac{V}{400} \right)^2 = \frac{1}{1.8} \qquad \text{or} \qquad V = \frac{400}{\sqrt{1.8}} = 298.1$$

(b) Currents are proportional to the applied voltage.

$$\therefore 6I_f \propto 400; \quad I \propto 298.1$$

$$\therefore I = 6 \times \frac{298.1}{400}, \quad I_f = \frac{6 \times 298.1 \times 30}{400} = 134.2 \text{ A}$$

(c) Here $K = \dfrac{298.1}{400}$

$$\text{Line current} = K^2 I_{sc} = \left(\frac{298.1}{400} \right)^2 \times 6 \times 30 = 100 \text{ A}$$

(d) We have seen that line current $= K^2 I_{sc}$

$$\text{Now, line current} = \text{full-load current } I_f$$

$$\therefore 30 = K^2 \times 6 \times 30 \therefore K^2 = \frac{1}{6}$$

No, using $\frac{T_{st}}{T_f} = K^2 \left(\frac{I_{sc}}{I_f}\right)^2 S_f$, we get $\frac{T_{st}}{T_f} = \frac{1}{6} \times \left(\frac{6I_f}{I_f}\right)^2 \times 0.05 = 0.3$

Here $N_s = 120 \times 50/6 = 1000$ rpm, N=950 rpm, $s_f = 50/1000 = 0.05$

$$\therefore T_{st} = 0.3T_f \quad \text{or 30\% F.L. torque}$$

Example 7.10: The full-load efficiency and power factor of a 12 kW, 440 V, 3-phase induction motor are 85% and 0.8 lag respectively. The blocked rotor line current is 45 A at 220 V. Calculate the ratio of starting to full-load current if the motor is provided with a star-delta starter. Neglect magnetizing current.

Solution:

Blocked rotor current with full voltage applied

$$I_{sc} = 45 \times \frac{440}{220} = 90 \text{ A}$$

Now, $\sqrt{3} \times 440 \times I_f \times 0.8 = 12000 / 0.85, \therefore I_f$

$$= 23.1 \text{ A}$$

In star-delta starter, $I_{st} = \frac{I_{sc}}{\sqrt{3}} = \frac{90}{\sqrt{3}} = 52$ A

$$\therefore \frac{I_{sc}}{I_f} = \frac{52}{23.1} = 2.256$$

Example 7.11: A 3-phase, 6-pole, 50 Hz, induction motor takes 60 A at full-load speed of 940 rpm and develops a torque of 150 N-m. The starting current at rated voltage is 300 A. What is the starting torque? If a star/delta starter is used, determine the starting torque and starting current.

Solution:

For direct-switching of induction motors

$\frac{T_{st}}{T_f} = K^2 \left(\frac{I_{sc}}{I_f}\right)^2 S_f$, Here, $I_{st} = I_{sc} = 300$ A (line value);

$I_f = 300$ A (line value),

$$S_f = \frac{(1000-940)}{1000} = 0.06\,; \quad T_f = 150\text{Nm}$$

$$\therefore T_{st} = 150\left(\frac{300}{60}\right)^2 \times 0.06 = 225 \text{ Nm.}$$

When star-delta starter is used

$$\text{Starting current} = 1/3 \times \text{starting current with direct starting} = 300/3 = 100 \text{ A}$$

$$\text{Starting torque} = 225/3 = 75 \text{ Nm}$$

Example 7.12: A 15 hp (11.2 kW), 3-phase, 6-pole, 50 Hz, 400 V, Δ-connected induction motor runs at 960 rpm on full-load. If it takes 86.4 A on direct starting, find the ratio of starting torque to full-load torque with a star-delta starter. Full-load efficiency and power factor are 88% and 0.85, respectively.

SOLUTION:

$$\text{Here, } I_{sc}/\text{phase} = \frac{86.4}{\sqrt{3}} \text{ A}$$

$$I_{st} \text{ per phase} = \frac{1}{\sqrt{3}} \cdot I_{sc} \text{ per phase} = \frac{86.4}{\sqrt{3}\times\sqrt{3}} = 28.8 \text{ A}$$

Full-load input line current may be found from

$$\sqrt{3} \times 440 \times I_L \times 0.85 = 11.2(10)^3/0.88$$

$$\therefore \therefore \text{Full-load } \mathrm{I}_L = 21.59\text{A}$$

$$N_s = 120 \times \frac{50}{6} \times 1000 \text{ rpm, N=950; } s_f = 0.05$$

$$\frac{T_{st}}{T_f} = K^2 \left(\frac{I_{sc}}{I_f}\right)^2 S_f = \left(\frac{28.8\sqrt{3}}{21.59}\right)^2 \times 0.05$$

or 26.7% F.L. torque. $\therefore T_{st} = 0.267\, T_f$

Example 7.13: A 480 V, 3-phase, 4-pole, 60 Hz, induction motor is running at a slip of 2.5%. Find:

(a) The speed of the magnetic fields in revolutions per minute
(b) The speed of the rotor in revolutions per minute
(c) The slip speed of the rotor
(d) The rotor frequency in hertz

SOLUTION:

(a) The speed of the magnetic fields is

$$N_s = 60 * \frac{60}{2} = 1800 \text{ rpm}$$

(b) The speed of the rotor is

$$N_m = (1 - s)N_s = 1755 \text{ rpm}$$

(c) The slip speed of the rotor is

$$N_{slip} = sN_s = 0.025 \times 1800 = 45 \text{ rpm}$$

(d) the rotor frequency

$$f_r = \frac{N_{slip} P}{120} = 1.5 \text{ Hz}$$

Example 7.14: A 3-phase, 60 Hz, induction motor runs at 715 r/min at no load and at 670 r/min at full load.

(a) How many poles does this motor have?
(b) What is the slip at rated load?
(c) What is the speed at one-quarter of the rated load?
(d) What is the rotor's electrical frequency at one-quarter of the rated load?

SOLUTION:

(a) This machine has ten poles, which produces a synchronous speed of

$$N_s = 60 * \frac{60}{2} = 720 \text{ rpm}$$

(b) The slip at rated load is

$$s = \frac{N_s - N}{N_s} = \frac{720 - 670}{720} = 6.94\%$$

(c) The motor is operating in the linear region of its torque-speed curve, so the slip at 1/4 load will be

$$N = (1 - s)N_s = (1\text{-}0.0174) \times 720\text{=}707 \text{ rpm}$$

(d) the electrical frequency at 1/4 load is

$$f_r = sf_e = 1.04 \text{ Hz}$$

Example 7.15: A 50 kW, 440 V, 50 Hz, 2-pole, induction motor has a slip of 6% when operating at full-load conditions. At full-load conditions, the friction and windage losses are 520 W, and the core losses are 500W. Find the following values for full-load conditions:

(a) The shaft speed N
(b) The output power in watts
(c) The load torque T_{load} in newton-meters
(d) The induced torque T_{ind} in newton-meters
(e) The rotor frequency in hertz

SOLUTION:

(a) The synchronous speed of this machine is

$$N_s = 60 * \frac{50}{1} = 3000 \text{ rpm}$$

Therefore, the shaft speed is

$$N = (1 - s)N_s = 2820 \text{ rpm}$$

(b) The output power in watts is 50 kW (stated in the problem).
(c) The load torque is

$$T_{load} = \frac{P_{out}}{\omega} = \frac{50,000}{N} = 169.3 \text{ Nm}$$

(e) The induced torque can be found as follows:

$$T_{ind} = \frac{P_{conv}}{\omega} = \frac{51,200}{2\pi * \frac{N}{60}} = 173.\text{Nm}$$

(f) The rotor frequency is

$$f_r = sf_e = 0.06 \times (50) = 3.0 \text{ Hz}$$

Example 7.16: A 3-phase, 60 Hz, 2-pole, induction motor runs at a no-load speed of 3,580 r/min and a full-load speed of 3,440 r/min. Calculate the slip and the electrical frequency of the rotor at no-load and full-load conditions. What is the speed regulation of this motor?

SOLUTION:

The synchronous speed of this machine is 3,600 rpm. The slip and electrical frequency at no-load conditions is

$$s = \frac{N_s - N}{N_s} = \frac{3600 - 3580}{3600} * 100\% = 0.56\%$$

$$f_r = sf_e = 0.0056 \times (60) = 0.33 \text{ Hz}$$

The slip and electrical frequency at full load conditions is

$$s = \frac{N_s - N}{N_s} = \frac{3600 - 3440}{3600} * 100\% = 4.44\%$$

The speed regulation is

$$SE = \frac{N_{nl} - N_{fl}}{N_{fl}} = \frac{3580 - 3440}{3440}$$

Problems

7.1 A 208 V, 4-pole, 60 Hz, Y-connected, wound-rotor induction motor is rated at 15 hp. Its equivalent circuit components are

$R_1 = 0.220\ \Omega, R_2 = 0.127\ \Omega, X_m = 15.0\ \Omega$
$X_1 = 0.430\ \Omega, X_2 = 0.430\ \Omega$
$P_{mech} = 300\ \text{W}, P_{misc} = 0, P_{core} = 200\ \text{W}$

For a slip of 0.05, find:

(a) The line current I_L
(b) The stator copper losses P_{SCL}
(c) The air-gap power P_g
(d) The power converted from electrical to mechanical form P_{conv}
(e) The induced torque T_{ind}
(f) The load torque T_{load}
(g) The overall machine efficiency
(h) The motor speed in revolutions per minute and radians per second.

7.2 A 440 V, 50 Hz, six-pole, Y-connected, induction motor is rated at 75 kW. The equivalent circuit parameters are:

$R_1 = 0.082\ \Omega, R_2 = 0.070\ \Omega, X_m = 7.2\ \Omega, X_1 = 0.19\ \Omega, X_2 = 0.18\ \Omega$
$P_{F\&W} = 1.3\ \text{kW}, P_{misc} = 150\ \text{W}\ P_{\text{core}} = 1.4\ \text{kW}$

For a slip of 0.04, find:

(a) The line current I_L
(b) The stator power factor
(c) The rotor power factor
(d) The stator copper losses P_{SCL}
(e) The air-gap power P_g
(f) The power converted from electrical to mechanical form P_{conv}
(g) The induced torque T_{ind}
(h) The load torque T_{load}
(i) The overall machine efficiency η.

7.3 A 208 V, 60 Hz, six-pole, Y-connected, 25-hp design class B induction motor is tested in the laboratory, with the following results:

No load: 208 V, 22.0 A, 1,200 W, 60 Hz
Locked rotor: 24.6 V, 64.5 A, 2,200 W, 15 Hz
DC test: 13.5 V, 64 A

Find the equivalent circuit of this motor and plot its torque-speed characteristic curve.

7.4 A 208-V, 4-pole, 10 hp, 60 Hz, Y-connected, 3-phase, induction motor develops its full-load induced torque at 3.8% slip when operating at 60 Hz and 208 V. The per-phase circuit model impedances of the motor are

$R_1 = 0.33\ \Omega, X_m = 16\ \Omega, X_1 = 0.42\ \Omega, X_2 = 0.42\ \Omega$

Mechanical, core, and stray losses may be neglected in this problem.

(a) Find the value of the rotor resistance R_2.
(b) Find T_{max}, $S_{\max}$, and the rotor speed at maximum torque for this motor.
(c) Find the starting torque of this motor.
(d) What code letter factor should be assigned to this motor?

8 Induction Motors Control

8.1 INTRODUCTION

The input impedance of a 3-phase induction motor at standstill is relatively small and of highly inductive reactance nature. When such a motor is directly connected to a 3-phase supply, an extremely high current tends to flow unless steps are taken to keep its value down to a reasonable level. So in general, an induction motor will be subjected to the most severe operating conditions during its starting and accelerating period. However, it is important that the starting current be minimized to such an extent that the line voltage drop does not cause any damage or objectionable change in the supply voltage, which would affect the operation of other equipment connected to the same source. Induction motors will usually take about 5 to 7 times the rated current at starting when the rated voltage is applied to their stators.

Therefore, in order to reduce the starting current by 50%, it is necessary that the starting voltage be lowered by this same percentage. One serious objection to the practice of using a reduced starting voltage is the large reduction in the starting torque (starting torque is proportional to the square of the impressed voltage).

8.2 STARTING OF INDUCTION MOTORS

There are different methods in use to start 3-phase induction motors. Some of the common methods are as follow:

8.2.1 Direct-On-Line Method

Direct-on-line (DOL) is the traditional and simplest method of motor starting, and most other methods are baselined against it. It is also often called across the line start as shown in Figure 8.1. This method is the direct connection of the terminal voltage to the motor stator with no additional components, and also for this reason is most economical in terms of installation cost and ease of use. It is also one of most reliable and robust methods. Of all the starting methods, it produces the highest inrush current, usually six to eight times the rated current, and the highest starting torque; and due to the high starting torque it has the shortest acceleration time. The DOL method is most commonly used for small motors relative to the size of the generation and system, due to the fact that the startup of a small motor will only have a low impact on the system, and in particular the voltage. Other drawbacks include the mechanical stress put on the motor's load and the low startup efficiency due to the high reactive power consumed at startup. This approach is typically not suitable for large motors.

DOI: 10.1201/9781003795841-8

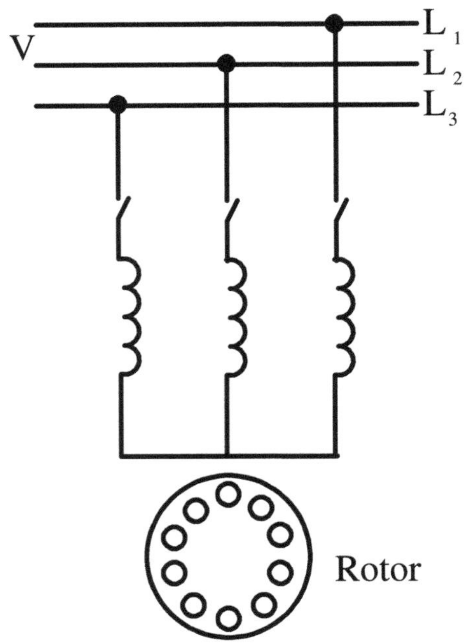

FIGURE 8.1 Direct on-line starting of induction motor.

8.2.2 AUTOTRANSFORMER

An autotransformer uses tap changes to reduce the low voltage as needed on the low side connected to the motor terminal. Therefore, the current can be reduced during startup; however, the torque is also reduced as the square of the voltage and needs to be taken into consideration to ensure enough torque is supplied during acceleration. The authors claim that 10% of the full load torque margin should be supplied at all points on the speed-torque characteristic curve. As the motor speeds up, the line is switched to the full voltage. Common taps range from 50% to 80% of the rated voltage.

IEEE Standard 399–1997 provides a table describing the currents at the typical 50%, 65%, and 80% of full voltage as 25%, 42%, and 64% of the current realized at full voltage during startup, respectively. A practical advantage of the autotransformer is the ability to provide different tap changes so that a wide range of applications, which vary in their starting torque and inrush current needs, can be performed. The star-delta starter is electrically equivalent to an autotransformer tapped at 57.7%. However, autotransformers have a higher cost than other conventional electromechanical starting methods.

Let x be the fractional percentage tappings used for an autotransformer to apply reduced voltage to the stator.

$$\begin{aligned} &\text{So if,} \quad I_{sc} = \text{Starting motor current at rated voltage} \\ &\text{and} \quad I_{st} = \text{Starting motor current with starter} \\ &\text{then} \quad I_{st} = x\, I_{sc} \quad \text{.....Motor side} \end{aligned} \tag{8.1}$$

But there exists a fixed ratio between starting current drawn from supply I_{st} (supply) and starting moor current I_{st} (motor) due to autotransformer, as shown in the Figure 8.2.

$$\text{Autotransformer ratio } x = I_{st}(\text{supply})/I_{st}(\text{motor})$$

$$I_{st}(\text{supply}) = x\ I_{st}(\text{motor})$$

Substituting I_{st} (motor) from equation (8.1), (8.2)

$$\therefore\ I_{st}(\text{supply}) = x\ .\ x\ I_{sc} = x^2\ I_{sc}$$

$$\text{Now } T_{st}\ \alpha\ I_{st}^{\ 2}(\text{motor})\ \alpha\ x^2 I_{sc}^{\ 2}$$

and $T_{F.L.}\ \alpha\ \left(I_{F.L.}\right)^2 / s_f$

$$\therefore \frac{T_{st}}{T_{F.L.}} = x^2 \left[\frac{I_{sc}}{I_{F.L.}}\right]^2 s_f \tag{8.3}$$

Note: Thus starting torque reduces by x^2 where x is the transformer ratio.

Example 8.1: A squirrel-cage induction motor has a full load slip of 5%. The motor starting current at rated voltage is 6 times its full load current. Find the tapping

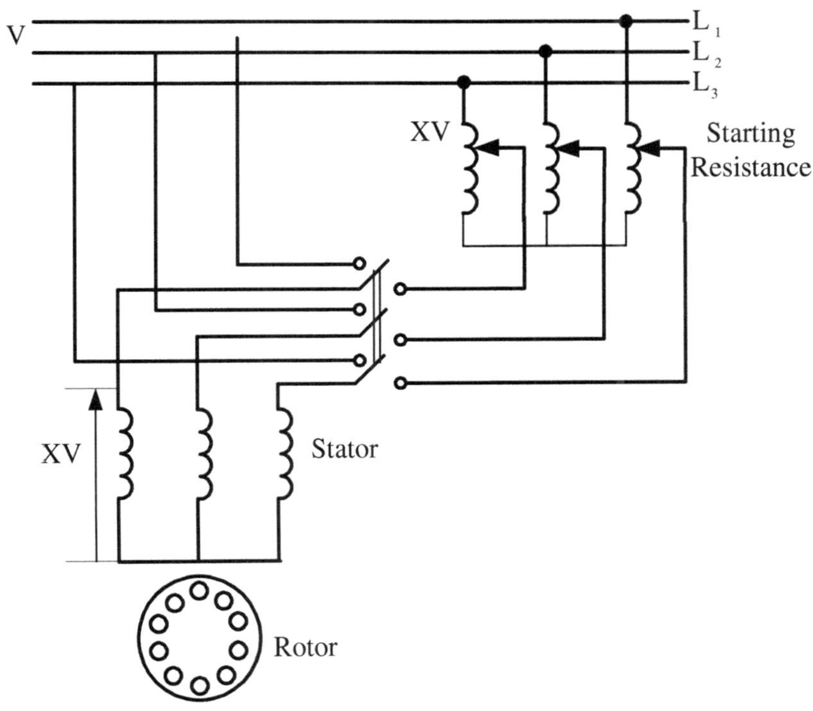

FIGURE 8.2 Autotransformer starting of induction motor.

on the autotransformer starter which would give full load torque at start. Whet would then be the supply starting current?

SOLUTION:

$$\text{Starting current at rated voltage} = I_{sc}$$

$$\therefore \quad I_{sc} = 6\, I_{F.L.} \quad \text{and } s_f = 5\% = 0.05$$

$$\text{Let} \quad x = \text{Tapping on autotransformer}$$

$$T_{F.L.} = T_{st}$$

$$\frac{T_{st}}{T_{F.L.}} = x^2 \left[\frac{I_{sc}}{I_{F.L.}} \right]^2 s_f$$

$$1 = x^2\ (6/1)^2 \times 0.05$$

$$x = 0.7453$$

Thus 74.53% tapping is required

$$\text{Now } I_{st}(\text{supply}) = x\ I_{st}(\text{motor}) = x\left(xI_{sc}\right) = x^2\ I_{sc}$$

$$= x^2\ (6) = 3.33\ I_{F.L.}$$

Thus supply starting current is 3.33 times the full load current.

8.2.3 PRIMARY RESISTOR

A switchable primary series resistor, shown in Figure 8.3, or reactor bank can be added at the motor terminals to limit the current or limit change in the current, respectively. The resistor bank will cause a drop in voltage across it reducing the current. The heat dissipated from the resistor also needs to be taken into consideration. Series resistor starting is usually only performed for small motors. When using a series reactor bank, it will oppose the inrush current initially and reduce the terminal voltage proportionally. The most advantageous characteristic of the series reactor starting is that the voltage increases over time as a function of the rate of change of the current without additional control. The added reactance will also further increase the starting reactive power and thus lower the starting efficiency. Switching transients will also occur if it is connected in an open-circuit.

$$\text{At start,} \quad s = 1, \quad T = T_{st} \quad \text{and } I_l = I_{st}$$

$$\therefore \quad T_{st} \ \alpha \ I_{st^2}$$

When stator resistance starter is used, the factory by which stator voltage reduces is say $x < 1$. The starting current is proportional to this factor x. So if I_{sc} is the normal current drawn under full rated voltage condition at start then,

$$I_{st} = x\ I_{sc}$$

$$\therefore \quad T_{st}\ \alpha \left(xI_{sc}\right)^2 \tag{8.4}$$

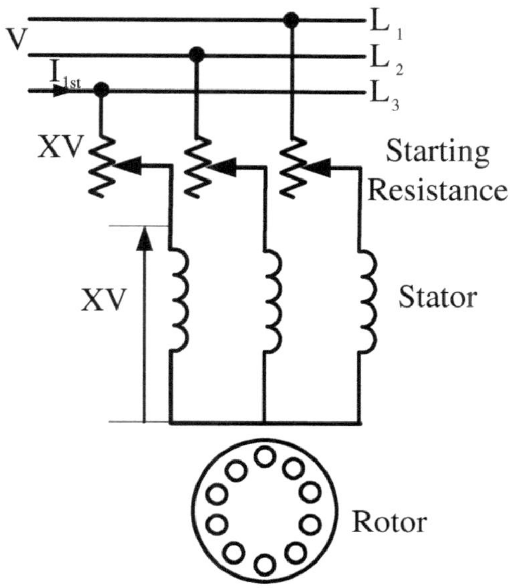

FIGURE 8.3 Stator resistance starting of induction motor.

$$\text{But} \quad T_{F.L.} \propto \left(I_{F.L.}\right)^2 / s_f \tag{8.5}$$

where s_f = Full load slip

Taking ratio of (8.4) and (8.5),

$$\therefore \frac{T_{st}}{T_{F.L.}} = x^2 \left[\frac{I_{sc}}{I_{F.L.}}\right]^2 s_f \tag{8.6}$$

Note: As $x < 1$, it can be seen that the starting torque reduces by the fraction x^2 due to the stator resistance starter.

8.2.4 Star-Delta

The star-delta (wye-delta) starting method controls whether the lead connections from the motor are configured in a star or delta electrical connection. The initial connection should be in the star pattern that results in a reduction of the line voltage by a factor of $1/\sqrt{3}$ (57.7%) to the motor and the current is reduced to 1/3 of the current at full voltage, but the starting torque is also reduced 1/3 to 1/5 of the DOL starting. The motor must be delta connected at rated voltage. The transition from star to delta transition usually occurs once nominal speed is reached but is sometimes performed as low as 50% of nominal speed. The star-delta method is usually only applied to low to medium voltage motors. The operation of the star-delta method is simple and rugged and is relatively cheap compared to other reduced voltage methods with only additional contactors added to the cost. However, the system cannot be modified once installed without considerable rework. Closed-circuit transitions can be performed to avoid open-circuit current surges as shown in Figure 8.4.

We have seen in case of autotransformer that if x is the factor by which the voltage is reduced then,

$$\therefore \frac{T_{st}}{T_{F.L.}} = x^2 \left[\frac{I_{sc}}{I_{F.L.}} \right]^2 s_f$$

Now the factor x in this type of starter is $1/\sqrt{3}$.

$$\therefore \frac{T_{st}}{T_{F.L.}} = \frac{1}{3} \left[\frac{I_{sc}}{I_{F.L.}} \right]^2 s_f \tag{8.7}$$

where I_{sc} = Starting phase current when delta connection with rated voltage
$I_{F.L.}$ = Full load phase current when delta connection.

8.2.5 Rotor Resistance Starting

This method allows external resistance to be connected to the rotor through slip rings and brushes. Initially, the rotor resistance is set to maximum and is then gradually decreased as the motor speed increases, until it becomes zero. The rotor impedance

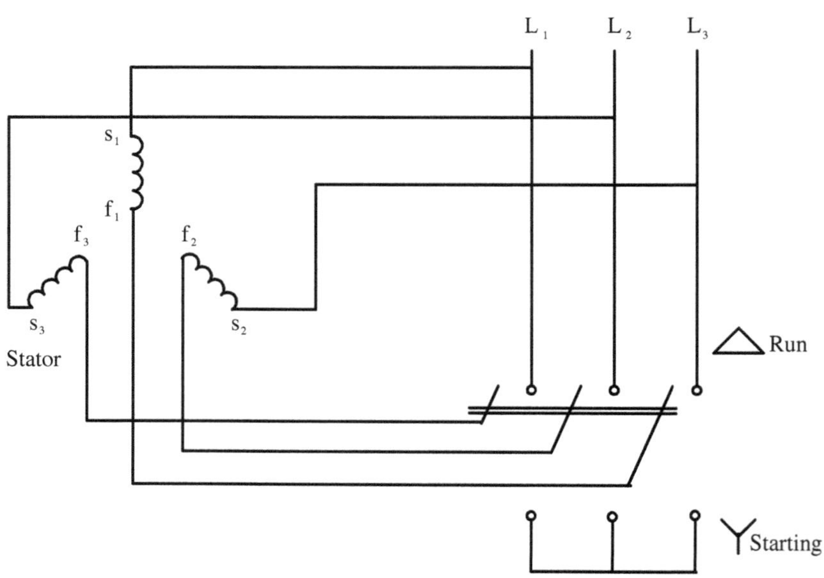

FIGURE 8.4 Star-delta starting of induction motor.

starting mechanism is usually very bulky and expensive when compared with other methods. It also has very high maintenance costs. Also, a considerable amount of heat is generated through the resistors when current runs through them. The starting frequency is also limited in this method. However, the rotor impedance method allows the motor to be started while on load. Figure 8.5 shows the connection of a 3-phase induction motor with rotor resistance starter.

8.2.6 Soft-Starter Techniques

A soft-starter is any solid-state electronic circuit-based device that manipulates the supply voltage prior to connecting to the motor terminals. Many different topologies for soft-starters exist. The soft-starter can be based on controlling different starting characteristics, including current, torque, and voltage, and can be easily adjusted based on the different loadings. Two basic types of soft-starters are the voltage ramping and current limiting types. The voltage ramping soft-starter is able to gradually increase the voltage from a preset level, as low as 0V, to the rated voltage; the result being a very smooth startup. The current limiting soft-starter senses the current at

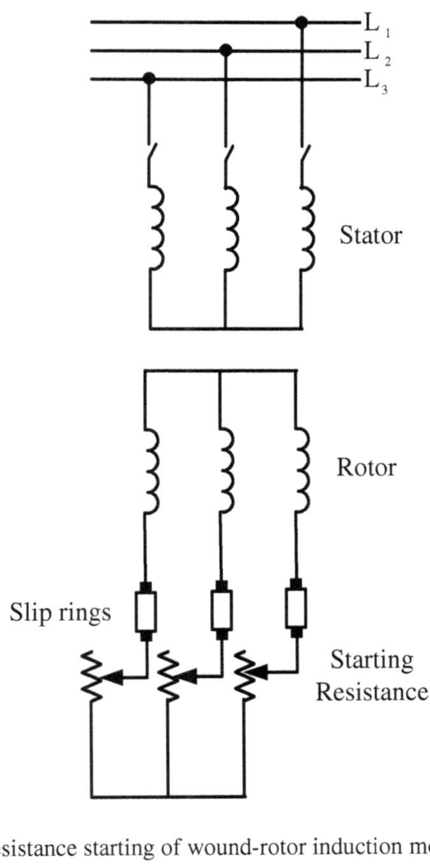

FIGURE 8.5 Stator resistance starting of wound-rotor induction motor.

the motor so that the fire angle can be controlled in a manner that the voltage is regulated to maintain the desired current.

Advantage of Soft Start

The various advantages of soft start are,

1. Absence of torque pulsation hence no jerks while starting.
2. Motor accelerates smoothly.
3. Due to use of electronic devices like thyristors the weight, size and cost is low.
4. The simpler control is possible.
5. The efficient operation is possible.

8.3 SPEED CONTROL OF 3-PHASE INDUCTION MOTOR

A 3-phase induction motor is practically constant speed motor provided that applied parameters (supply, frequency etc.) remains unchanged. The speed can be varied according to requirement. The speed of induction motor depends on various factors, like supplied voltage, frequency, and number of poles. There are different methods for controlling the speed from stator side and rotor side. These speed control methods are explained below.

8.3.1 Changing the Applied Frequency

We know that the synchronous speed of induction motor is directly proportional to the frequency of the voltage applied to its terminals. Thus speed can be varied by changing supply frequency. This will require a variable frequency, 3-phase source capable of supplying sufficient power to operate the motor. As we increase supply frequency, speed of the motor also increases and vice versa. This method is being used to some extent in electrically driven ships.

8.3.2 Pole Changing

It must be realized that the poles of an induction motor are created by proper arrangement of the stator conductors. It is also clear that for induction motors, there are no physical poles projecting from the stator as in the case of DC machines. Therefore, a given stator could be wound for two, four, six, or eight poles merely by interchanging the end connections of the stator coils.

A two- speed motor results when a two separate winding motor can be identified. The disadvantages of this method are:

(a) A large stator size is required in comparison of a single-speed motor.
(b) Excessive overheating is developed due to the induced EMFs in the idle winding when the other winding is connected to the supply. However, this effect can be eliminated by keeping the idle winding open circuited.

8.3.3 Variation of Rotor Resistance

The Wound-Rotor Induction Motor (WRIM) has the ability to control its effective rotor resistance by the use of an external variable resistance. In this case, the operating speed can be smoothly and continuously raised or lowered by changing the external rotor resistance. However, the rotor copper loss will high especially at large slips, so that this method of speed control can be efficient.

8.4 SQUIRREL-CAGE INDUCTION MOTOR

The 3-phase squirrel-cage induction motor can, and many times does, have the same armature (stator) winding as the 3-phase synchronous motor. As in the synchronous motor, applying 3-phase currents to the armature creates a synchronously rotating magnetic field. The induction motor rotor is a completely short-circuited conductive cage. Figures 8.6 illustrates the rotor construction.

The rotor receives its excitation by induction from the armature field. Hence, the induction machine is a doubly excited machine in the same sense as the synchronous and DC machines. The basic principle of operation is described by Faraday's Law. If we assume that the machine rotor is at a standstill and the armature is excited, then the armature produced rotating field is moving with respect to the rotor. In fact, the relative speed between the rotating field and the rotor is synchronous speed. For this condition, the rotating field induces a large voltage in the rotor bars. The large voltage causes a large current in the squirrel-case which, in turn, creates a magnetic field in the rotor. The rotor magnetic field interacts with the armature magnetic field, and a torque is produced. If the produced torque is larger than any load torque, the rotor begins to turn. As the rotor accelerates, the speed difference between the rotor and the armature field is reduced. This reduced speed difference (or slip) causes the

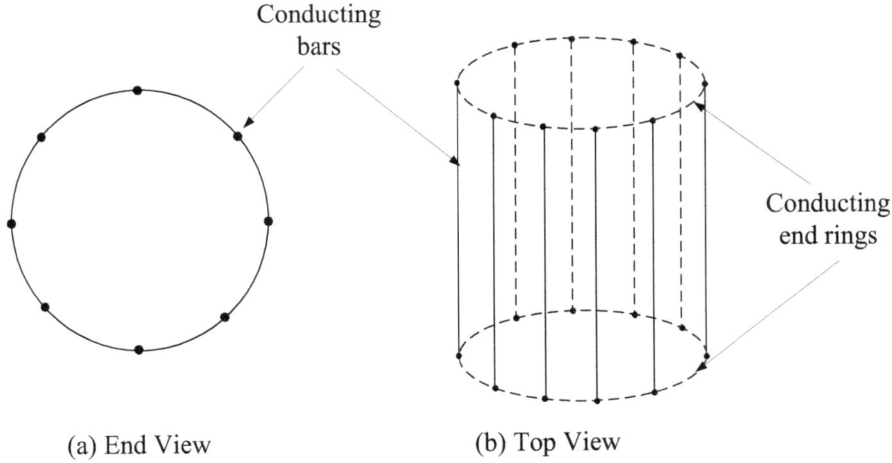

FIGURE 8.6 Squirrel-cage rotor.

induced rotor voltage to be reduced, the rotor current to be reduced, the rotor flux to be reduced, and the torque produced by the machine to be reduced. Eventually, the torque produced by the motor equals the torque demanded by the load, and the motor settles to an equilibrium rotor speed. This equilibrium rotor speed must be less than synchronous speed since there must be a slip to produce torque.

The frequency-dependent nature of the rotor impedances causes the torque versus speed characteristic of the induction motor to be quite non-linear. Figure 8.7 shows a typical characteristic.

To draw the vector diagram of induced EMF in the rotor bars, the number of bars is assumed to be 12 and the number of poles is 2. The 3-phase current in the stator winding generate a rotating MMF in the air gap which rotate in with a speed equal to the synchronous speed. This rotating MMF induces EMF in the short-circuited rotor bars. The EMF waveform and distribution is shown in Figure 8.8. It is obvious that the electrical angle α between the bar EMF is equal to the 360/S which is 30° which means that the number of phases is 12.

If the number of poles is increased to four, the electrical angle will be 60°. In this case, the EMF in two slots will be in parallel as show in Figure 8.9 making 6 phases. The EMF induced in bars repeat itself after 360° or 2-pole pitches and the conductors appear as if they are connected in parallel as shown in Figure 8.10. Therefore, the number of phases is equal to the number of rotor bars divided by the number of pole pairs.

$$m_2 = \frac{S_2}{P} \tag{8.8}$$

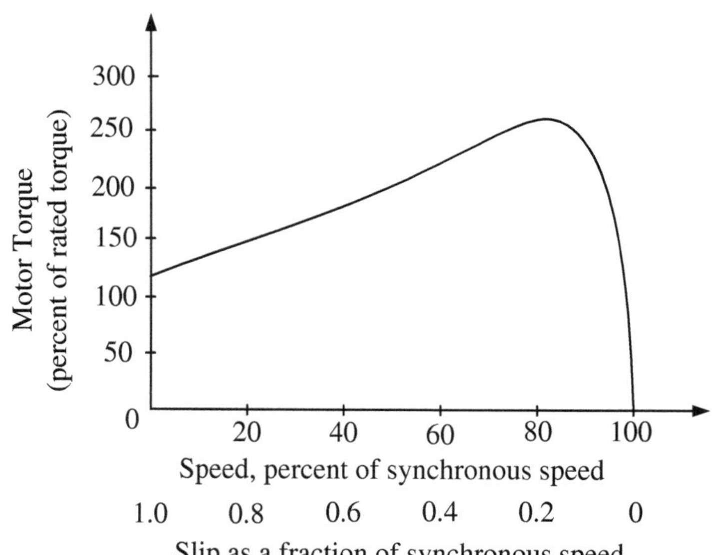

FIGURE 8.7 Torque-speed characteristics of induction motor.

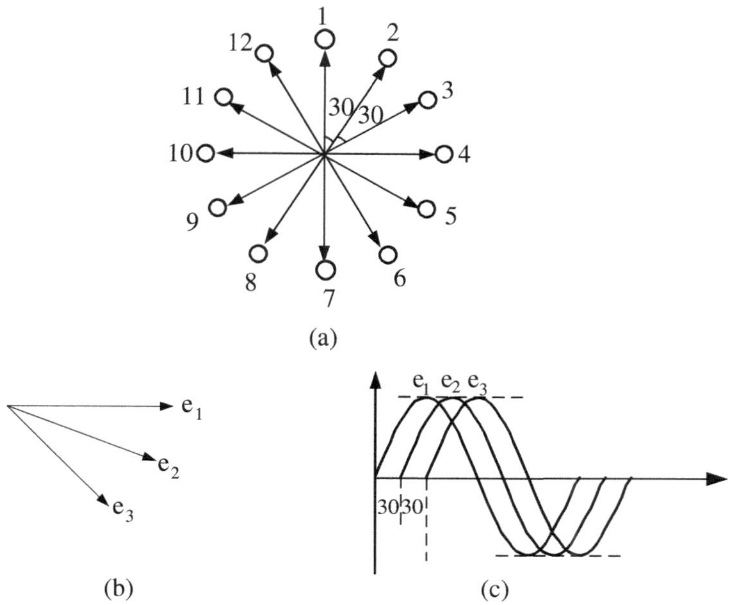

FIGURE 8.8 EMF in rotor bars.

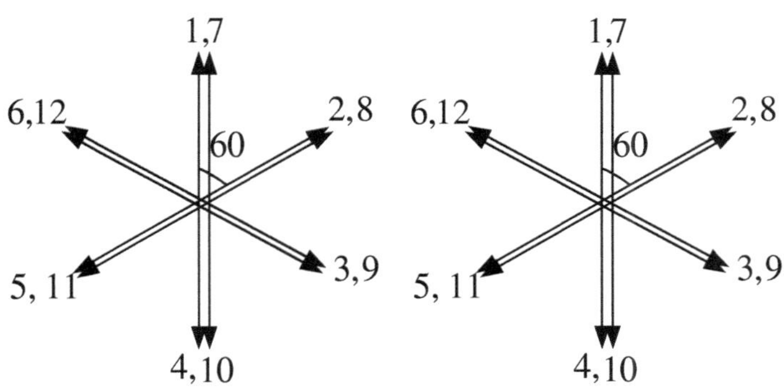

FIGURE 8.9 EMF in 4-pole, 12 bar SCIM.

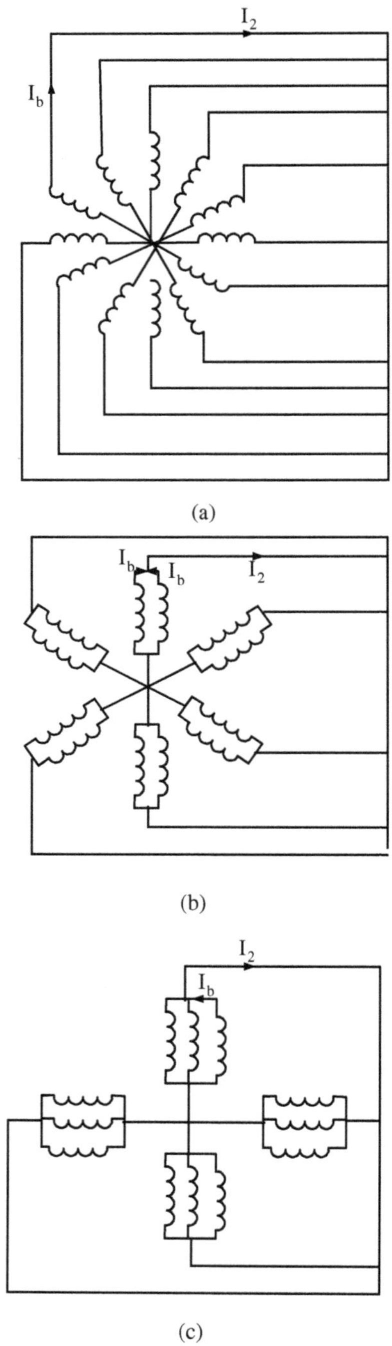

FIGURE 8.10 The effect of changing the number of poles in the rotor current.

and the rotor current is expressed as follow:

$$I_2 = PI_b \tag{8.9}$$

where m_2 is number of rotor phases, I_b is the bar current, I_2 is the rotor current, and P is number of pole pair.

Rotor Bar Current

Bar current in the rotor of squirrel-cage induction motor may be expressed as:

$$I_2' = I_2 \frac{T_2 K_{w2}}{T_1 K_{w1}} \tag{8.10}$$

where K_{w1} is the stator winding factor, K_{w1} is the rotor winding factor, T_1 is the stator number of turns, and T_2 is the stator number of turns.

The MMF of the stator should be equal to the rotor MMF as:

$$m_1 I_2' T_1 K_{w1} = m_2 I_2 T_2 K_{w2} \tag{8.11}$$

For squirrel-cage induction motor, $m_1 = 3$, $m_2 = S_2 / P$, and $T_2 = \frac{1}{2}$. Since the rotor is squirrel-cage, the winding factor $K_{w2} = 1$ as there are no chording or distribution factor.

$$\therefore \quad I_2 = I_2' \frac{m_1}{m_2} \frac{T_1}{T_2} \frac{K_{w1}}{K_{w2}}$$

$$\therefore \quad I_2 = \frac{3}{S_2 / P} \frac{T_1}{1/2} \frac{K_{w1}}{1} I_2' = \frac{6 T_1 K_{w1}}{P S_2} I_2' \tag{8.12}$$

By substituting from (8.9) in (8.12):

$$\therefore \quad P I_b = \frac{6 T_1 K_{w1}}{S_2 / P} I_2'$$

$$\therefore \quad I_b = \frac{6 T_1 K_{w1}}{S_2} I_2' \tag{8.13}$$

This equation may be re-written as:

$$\therefore \quad I_b = \frac{6 T_1 K_{w1}}{S_2} I_1 \cos\theta = 0.85 \frac{6 T_1 K_{w1}}{S_2} I_1 \tag{8.14}$$

Rotor End-Ring Current

All the rotor bars are short circuited by connecting them to the end rings at both the end rings. The rotating magnetic field produced will induce an EMF in the rotor bars which will be sinusoidal over one pole pitch. As the rotor is a short-circuited body, there will be current flow because of this EMF induced. The distribution of current and end rings are as shown in Figure 8.11. Referring to the figure considering the bars under one pole pitch, half of the number of bars and the end ring carry the current in one direction and the other half in the opposite direction. Thus, the maximum end ring current may be taken as the sum of the average current in half of the number of bars under one pole.

As discussed earlier, the currents of rotor bars are equal in magnitude and shifted by the bar angle γ as shown in Figure 8.12.

$$\therefore \quad |I_{EA}| = |I_{AB}| = |I_{BC}| = |I_e|$$

The vector component of the end ring current is as follow:

$$\bar{I}_{AB} = \bar{I}_{EA} + \bar{I}_{b1}, \bar{I}_{BC} = \bar{I}_{AB} + \bar{I}_{b2} \tag{8.15}$$

By neglecting the leakage reactance of the rotor winding, the bar currents are in phase with the bar EMF as the bar EMFs are shifted by angle γ from each other. From Figure 8.12, the eddy current may be expressed as:

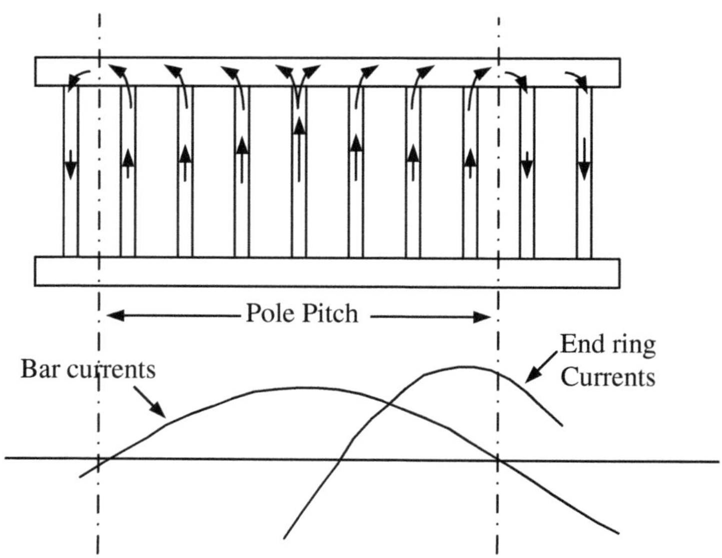

FIGURE 8.11 Currents in cage rotor bars and end ring.

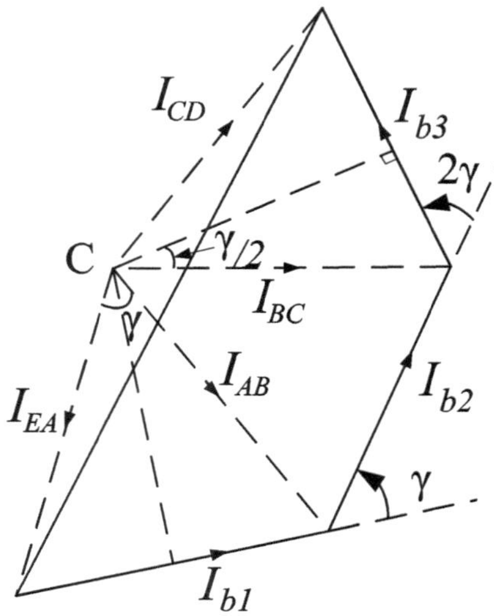

FIGURE 8.12 Triangle of currents.

$$I_e = \frac{\frac{I_b}{2}}{\sin\frac{\gamma}{2}} = \frac{I_b}{2\sin\frac{\gamma}{2}} \tag{8.16}$$

where γ is the rotor angle and given as $\gamma = \frac{360}{S_2}P$

$$I_e = \frac{I_b}{2\sin\frac{360}{S_2}P} \tag{8.17}$$

8.5 SPECIAL ROTOR CONSTRUCTIONS

In case of squirrel-cage induction motor, there is no provision made for adding external resistance. If the resistance is designed in such a way that it gives better running performance, then it has high starting current and consequently low starting torque. This is major disadvantage of squirrel-cage induction motor although it is having the other qualities of low cost, ruggedness and maintenance free operation. Thus the designer had found different ways of improving the starting performance of the motor without affecting the running performance of the motor.

In squirrel-cage induction motor, high starting torque can be obtained by the use of deep bar or double cafe rotors. Both these types of rotors make use of skin effect in which distribution of current is not uniform, but the alternating current has the tendency to concentrate near the surface of the conductor. Due to this effect, effective area of cross section of the conductor is reduced and hence resistance of the conductor is increased when carrying alternating current.

The solid conductor can be considered to be makeup of large number of strands each carrying a small part of current. The inductance of each strand will vary according to the position. The strands in proximity of the center are surrounded greater magnetic flux and has greater inductance than near the surface. Due to high reactance at the center, the alternating current flows near the surface of the conductor. The skin effects depend upon nature of material, diameter of wire, shape of wire and frequency.

Thus the current in the rotor during starting is having the frequency of supply. While under running condition the frequency of rotor current reduces to slip frequency. This variation in frequency changes the rotor resistance as it depends on skin effect. During starting, it gives high resistance whereas it gives low resistance during running condition which is desirable. Thus the variation in rotor resistance can be achieved by deep bar or double cage construction of rotor and induction motor. Both these types of construction make use of skin effect phenomenon.

8.5.1 Deep Bar Rotor Construction

There is no constructional difference between stator of deep bar motor and that of ordinary induction motor. The rotor consists of deep bars, short circuited by two end rings one on each side. The deep and narrow rotor bar of rectangular cross section is shown in the Figure 8.13a. The other rotor bar shapes are shown in the Figure 8.13b. The magnetic leakage flux lines are shown by dotted lines, now consider that the bar consists of many number of layers of different depths. The top and bottom layers are shown in the Figure 8.13.

The leakage inductance of the bottom strips is greater than that of top strips as more flux links with bottom strip compared to top strip. All the strips are in parallel electrically. The bottom strip has greater leakage inductance than the top strip. During starting the rotor frequency is equal to the stator frequency and hence leakage reactance of bottom strip is largest and current in it is least. The top strip has low leakage reactance and current in it is large. Thus, the current in low reactance top strip will be greater than that in high reactance lower strip and the current will be forced towards the top of the slot and phase of current in upper strip will lead that of the current in lower one. Thus there is non-uniform distribution of current which is shown in the Figure 8.14. Due to this non-uniform distribution of current, and use to skin effect, effective are of cross-section decreases. Hence, rotor resistance increases resulting i high starting torque.

As leakage reactance is proportional to frequency, the non-uniform distribution of current depends upon the rotor frequency. The Figure 8.14 shows a curve indicating AC effective resistance to DC resistance with change in frequency for a copper bar of 2.5 cm deep. The skin effect is maximum when rotor is at standstill.

With the increase in rotor speed, the rotor frequency decreases, and skin effect also decreases. The reactances of different strips at this low frequency become almost equal and the current density over the conductor cross section becomes uniform so its AC resistance is equal to DC resistance. Thus, with deep bar rotor has a low starting current with high starting torque without affecting running performance of motor. The net reactance of deep bar rotor at standstill is higher than that in a

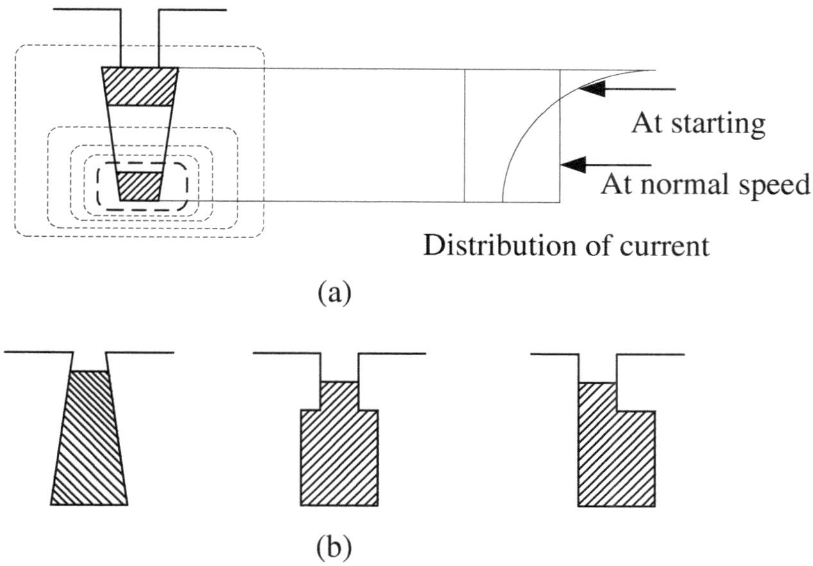

FIGURE 8.13 Types of slots in induction motor.

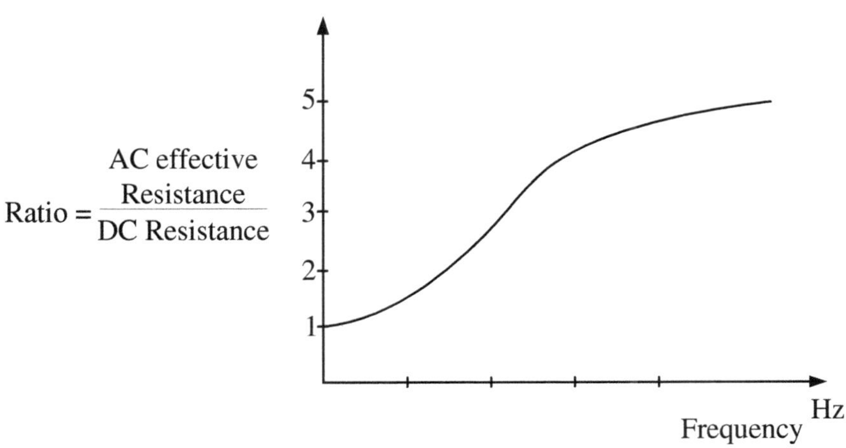

FIGURE 8.14 The effect of frequency changes on the ac resistance.

normal bar design, the breakdown or pull out torque in deep bar rotor is lower. The torque-slip characteristics of deep bar motor and normal induction motor is shown in the Figure 8.15.

The equivalent circuit of induction motor is applicable to deep bar rotor also wherein proper value of r_2' and x_2' must be determined for satisfactory running performance. During starting, their values should correspond to effective value at stator frequency. During running, their values should correspond to their effective values at low rotor frequency.

8.5.2 Double Cage Rotor Construction

This is another way of obtaining improved starting performance without affecting its running performance. Though it is more expensive, it gives better performance than deep bar rotor construction.

The stator of double cage rotor induction motor is same as that of ordinary induction motor whereas its rotor consists of two cages or two layers of bars short circuited by end rings since the upper cage is having smaller cross-sectional area than the lower cage, the upper cage is having higher resistance than that of lower cage. With equal cross-sectional areas of two cages, the upper cage is made up of high resistance material like brass, aluminum, bronze, etc. and the lower cage is made up of low resistance material like copper. The upper cage and lower cage are separated by a narrow slit or constriction. This is shown in the Figure 8.16.

The slot leakage flux pattern is also shown in the Figure 8.16 for the double cage rotor. Similar to deep bar rotor construction the rotor bars in the upper cage have less leakage flux linkage and therefor has lower reactance. The dimension of air construction controls the self-leakage flux linking upper and lower bars. If air constriction would have been absent, then the main flux would return via iron path between the

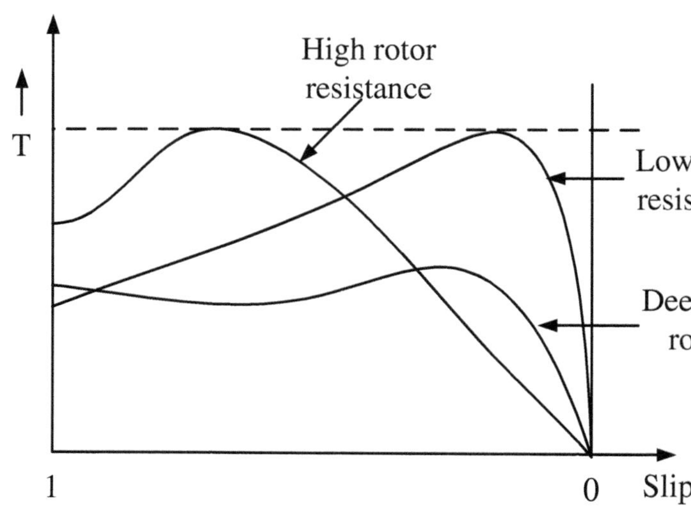

FIGURE 8.15 Torque slip characteristics of deep bar rotor.

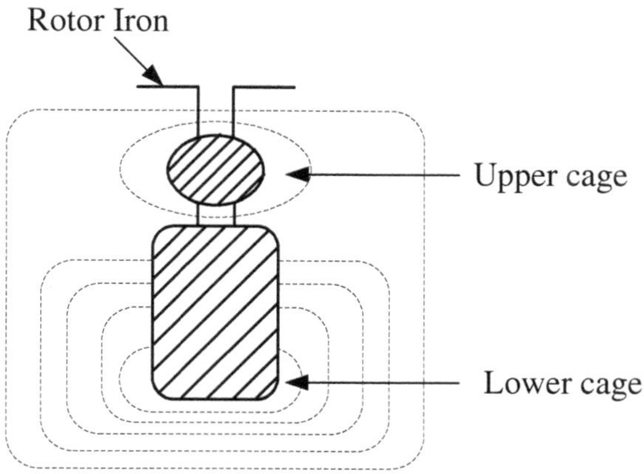

FIGURE 8.16 Double cage rotor construction.

two slots and thereby missing the bars in the lower cages which will not contribute to production of torque in that case. Hence, it can be seen that the upper cage has high resistance and low reactance, whereas the lower cage has low resistance and high reactance.

During starting, the rotor frequency is same as stator frequency or supply frequency. The division of rotor current in upper and lower cage is inversely proportional to their leakage impedances. At the time of starting the leakage reactance of lower cage is very high and consequently its leakage impedance is several times greater than that of upper cage whose leakage reactance is small. Hence, most rotor current flows in upper cage having lower leakage impedance. The upper cage having high resistance sharing the rotor current results in low starting current at improved power factor giving high starting torque.

When rotor speeds up, the rotor frequency decreases which decreases the leakage reactance of lower cage. At normal operating speed the reactance difference between the two cages is negligibly small. Hence the division of rotor current in this case is mainly decided by the resistances of the two cages. As resistance of upper cage is very high most of the current flows through the lower cage giving excellent operating characteristics under running condition. It can be noted that starting current is confined mainly with upper cage so if there is frequent starting of motor then it would cause overheating and burning of upper cage.

The torque-slip characteristics of double-cage induction motor are shown in the Figure 8.17. Another type of double cage rotor construction is also possible which is shown in the Figure 8.18. The slot-leakage flux pattern for this type of construction is also shown.

The approximate equivalent circuit of double cage rotor induction motor is shown in the Figure 8.19. Though the two cages are somewhat coupled magnetically, they can be treated as independent for simplicity and it gives approximately same results. The two cages are assumed to be parallel while drawing the equivalent circuit.

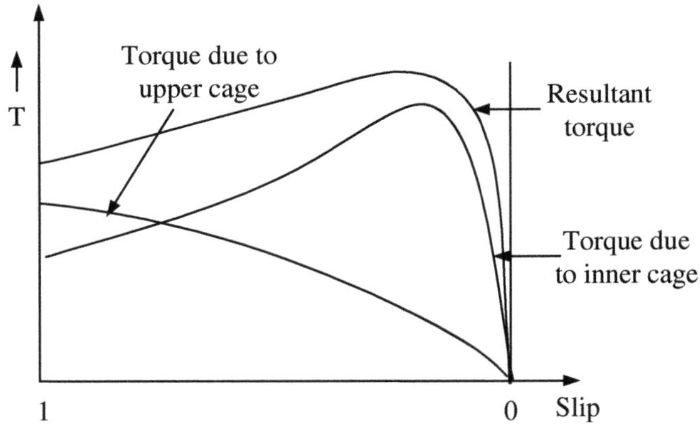

FIGURE 8.17 Torque slip characteristics of double cage induction motor.

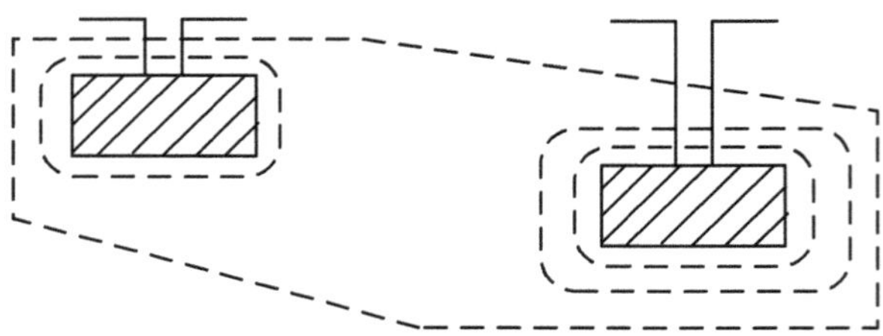

FIGURE 8.18 Double-cage rotor and leakage flux.

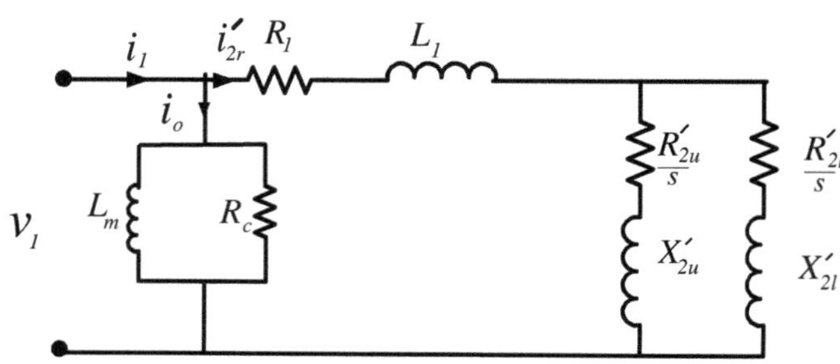

FIGURE 8.19 Equivalent circuit of double-cage induction motor.

I_{2ru}' and I_{2r}' are the currents in the upper and lower cages, respectively, referred to the stator; R_{2u}' and R_{2e}' are the resistance of upper and lower cages referred to the stator, whereas and are leakage reactances of the two cages referred to the stator of the motor.

8.6 CIRCLE DIAGRAM

This is a very useful tool to determine the performance of a motor with the help of only no-load and short-circuit test results. In slip-ring motors, it also helps to determine the external resistance required in the rotor circuit to control the speed of the motor and achieve the desired operating performance. The concept behind this diagram is that the locus of the rotor and the stator currents is a circle. Consider the equivalent circuit of an induction motor as shown in Figure 8.20.

where

R_1= stator resistance
X_1= stator reactance
R_2'= rotor resistance referred to stator
X_2'= rotor reactance referred to stator
I_{nl}= no-load current

8.6.1 Drawing the Circle Diagram

1. Take V_l on the vertical axis and draw I_0 at an angle φ_0 obtained from the no-load test.
2. From a short-circuit test draw the start-up current I_{st} at an angle φ_{st}.
3. Join AB and determine the center C and draw the circle as shown in Figure 8.21. The diameter of the circle can also be determined by:

$$AD = \frac{V_{xx}}{X_1 + X_2'} \tag{8.18}$$

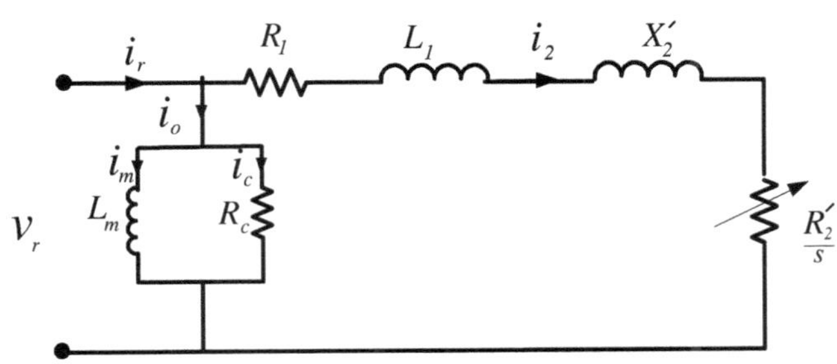

FIGURE 8.20 The per-phase approximate equivalent circuit of an induction motor.

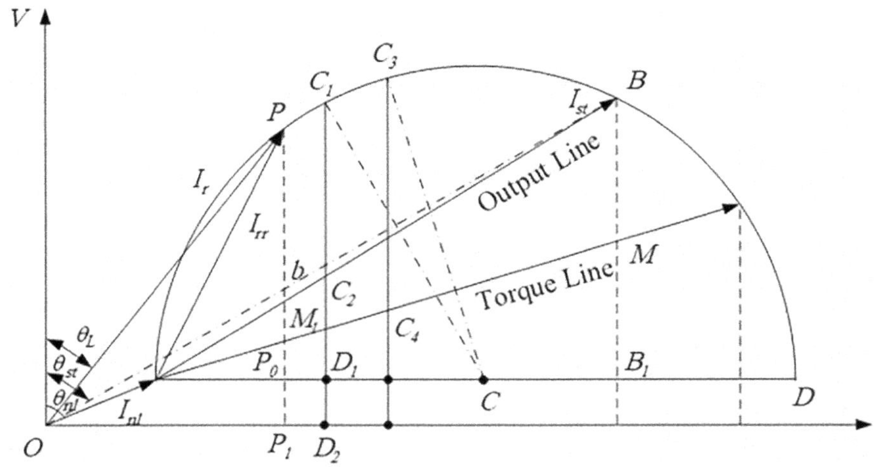

FIGURE 8.21 Construction of IM circle diagram.

4. BB′ will determine the locked rotor torque and power loss while the rotor is locked.
5. Divide BB′ at M in the ratio of $R_2' : R_1$ and join AM.
6. At load current I_r, that is, OP, the rotor current I_{rr} is AP, PP_1 will determine the power input per phase and the performance of the motor as follows:
   (i) Power input = $3 \cdot V_l \cdot PP_1$ watts
   (ii) Core and friction loss = $3 \cdot V_l \cdot D_1D_2$ watts
   (iii) Stator copper loss = $3 \cdot V_l \cdot M_1P_0$ watts
   (iv) Rotor copper loss = $3 \cdot V_l \cdot bM_1$ watts
   (v) Motor output = $3 \cdot V_l \cdot P_b$ watts AB is known as the output line since the output is measured above this.
   (vi) Running torque, $T = 3 \cdot V \cdot PM_1$ synchronous watts and since $3 \cdot V \cdot PM_1 = T\,N_r/0.974$

$$\therefore T = \frac{3 \cdot V_1 \cdot 0.974 \cdot PM_1}{N_r} \text{ kgm} \tag{8.19}$$

   This is the maximum torque that can be developed by the rotor during a run, but the useful torque will be in accordance with output,

$$\text{i.e. } \frac{3 \cdot V_1 \cdot 0.974 \cdot P_b}{N_r} \text{ kgm}$$

   Since the maximum torque is measured by line AM, it is known as the torque line.

   (vii) Starting torque or short-circuit torque $T_{st} = \dfrac{3 \cdot V_1 \cdot 0.974 \cdot PM}{N_r}$
   (viii) Full-load slip, $s = \dfrac{bM_1}{PM_1}$

8.6.2 Maximum Quantities

It will now be shown from the circle diagram (Figure 8.22) that the maximum values occur at the positions stated below:

(i) Maximum Output: It occurs at point M where the tangent is parallel to output line O'A. Point M may be located by drawing a line CM from point C such that it is perpendicular to the output line O'A. Maximum output is represented by the vertical MP.

(ii) Maximum Torque or Rotor Input: It occurs at point N where the tangent is parallel to torque line O'E. Again, point N may be found by drawing CN perpendicular to the torque line. Its value is represented by NQ. Maximum torque is also known as stalling or pull-out torque.

(iii) Maximum Input Power It occurs at the highest point of the circle, that is, at point R where the tangent to the circle is horizontal. It is proportional to RS. As the point R is beyond the point of maximum torque, the induction motor will be unstable here.

However, the maximum input is a measure of the size of the circle and is an indication of the ability of the motor to carry short- time over-loads. Generally, RS is twice or thrice the motor input at rated load.

Example 8.1: A 3-phase, 400 V induction motor gave the following test readings; No-load: 400 V,1250 W, 9 A, Short-circuit: 150 V, 4 kW, 38 A. Draw the circle diagram. If the normal rating is 14.9 kW, find from the circle diagram, the full-load value of current, p.f. and slip.

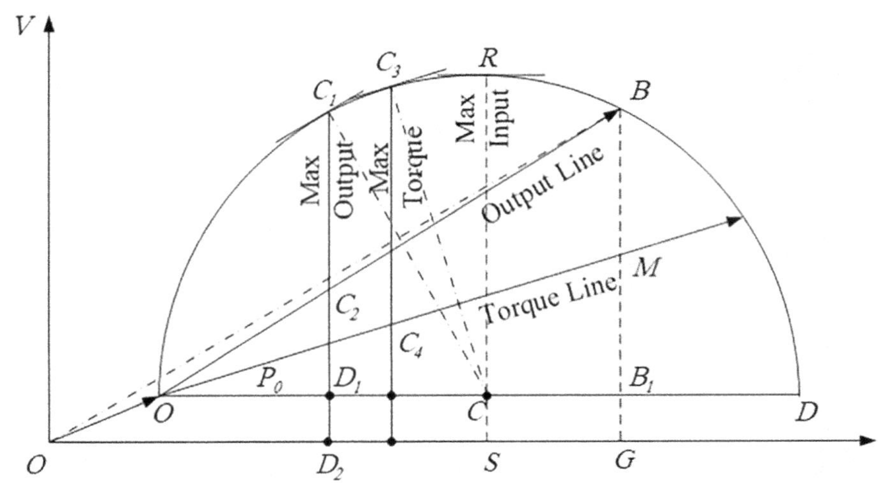

FIGURE 8.22 Circle Diagram Maximum quantities

SOLUTION:

$$\cos\theta_{nl} = \frac{1250}{\sqrt{3}\times 400\times 9} = 0.2004;\quad \theta_{nl} = 78.5^0$$

Short-circuit current with normal voltage is $I_{st} = 38\,(400/150) = 101.3$ A. Power taken would be $= 4000\,(400/150)^2 = 28,440$ W. In the above figure, OO′ represents I_{nl} of 9 A. If current scale is 1 cm = 5 A, then vector OO′ = 9/5 = 1.8 cm and is drawn at an angle of $\theta_{nl} = 78.5°$ with the vertical OV (which represents voltage). Similarly, OA represents I_{st} (S.C. current with normal voltage applied) equal to 101.3 A.

It measures 101.3/5 = 20.26 cm and is drawn at an angle of 66.1°, with the vertical OV. Line O′D is drawn parallel to OX. NC is the right-angle bisector of O′A. The semi-circle O′AD is drawn with C as the center. This semi-circle is the locus of the current vector for all load conditions from no-load to short-circuit. Now, AF represents 28,440 W and measures 8.1 cm. Hence, power scale becomes: 1 cm = 28,440/8.1 = 3,510 W. Now, full-load motor output $= 14.9(10)^3 = 14,900$ W. According to the above calculated power scale, the intercept between the semi-circle and output line O′A should measure = 14,900/3510 = 4.25 cm. For locating full-load point P, BA is extended. AS is made equal to 4.25 cm and SP is drawn parallel to output line O′A. PL is perpendicular to OX. Line current = OP = 6 cm = 6×5 = 30 A; $\theta = 30°$ (by measurement) p.f. = cos 30° = 0.886 (or cos θ = PL/OP = 5.2/6 = 0.865).

$$\text{Now, slip} = \frac{\text{rotor Cu loss}}{\text{rotor input}}$$

In the above figure, EK represents rotor Cu loss and PK represents rotor input.

$$\therefore \text{slip} = \frac{\text{EK}}{\text{PK}} = \frac{0.3}{4.5} = 0.067 \text{ or } 6.7\%.$$

Example 8.2 Draw the circle diagram for a 3.73 kW, 200 V, 50 Hz, 4-pole, 3-phase, star-connected induction motor from the following test data:

No-load: Line voltage 200 V, line current 5 A; total input 350 W

Blocked rotor: Line voltage 100 V, line current 26 A; total input 1,700 W

Estimate from the diagram for full-load condition, the line current, power factor and also the maximum torque in terms of the full-load torque. The rotor Cu loss at standstill is half the total Cu loss.

SOLUTION:

$$\text{From No-load}: I_{nl} = 5\text{A}, \cos\theta_{nl} = \frac{350}{\sqrt{3}*200*9} = 0.202;\quad \theta_{nl} = 78.15^0$$

$$\text{From Blocked rotor}: \cos\theta_s = \frac{1700}{\sqrt{3}*100*26} = 0.378;\quad \theta_s = 67.42^0$$

Short-circuit current with normal voltage, $I_{st} = 26 \times 200 / 100 = 52$ A.

Short-circuit / blocked rotor input with normal voltage $= 1700\,(52/26)^2 = 6{,}800$ W.

In the circle diagram, voltage is represented along OV which is drawn perpendicular to OX. Current scale is $1 \text{ cm} = 2\text{A}$. Line OA is drawn at an angle of $\theta_{nl} = 78^\circ 15'$ with OV and 2.5 cm in length. Line AX′ is drawn parallel to OX. Line OB represents short-circuit current with normal voltage, that is, 52 A and measures $52/2 = 26$ cm. AB represents output line. Perpendicular bisector of AB is drawn to locate the center C of the circle. With C as center and radius = CA, a circle is drawn which passes through points A and B. From point B, a perpendicular is drawn to the base. BD represents total input of 6,800 W for blocked rotor test. Out of this, ED represents no-load loss of 350 W and BE represents $6{,}800 - 350 = 6{,}450$ W. Now BD = 9.8 cm and represents 6,800 W.

$$\therefore \text{power scale} = 6{,}800 / 9.8 = 700 \text{ watt / cm or } 1 \text{ cm} = 700 \text{ W}$$

BE which represents total copper loss in rotor and stator is bisected at point T to separate the two losses. AT represents torque line.

Now, motor output $= 3{,}730$ watt. It will be represented by a line $= 3{,}730/700 = 5.33$ cm

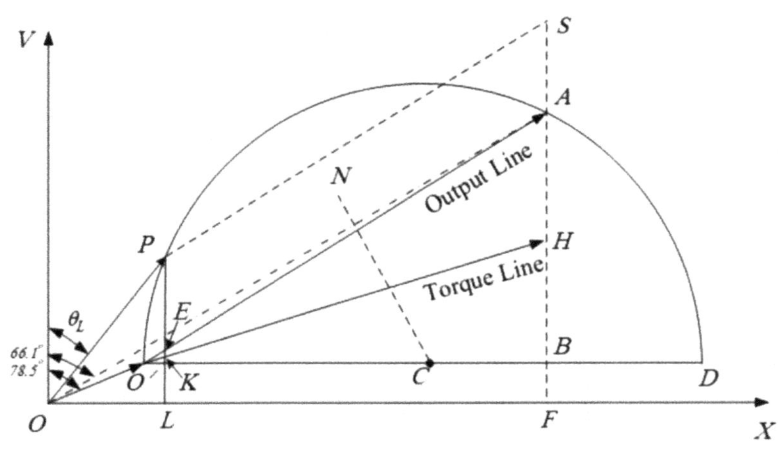

The output point P on the circle is located thus:

DB is extended and BR is cut = 5.33 cm. Line RP is drawn parallel to output line AB and cuts the circle at point P. Perpendicular PS is drawn and P is joined to origin O. Point M corresponding to maximum torque is obtained thus:

From center C, a line CM is drawn such that it is perpendicular to torque line AT. It cuts the circle at M which is the required point. Point M could also have been located by drawing a line parallel to the torque line. MK is drawn vertical and it represents maximum torque.

Now, in the circle diagram, OP = line current on full-load = 7.6 cm. Hence, OP represents $7.6 \times 2 = 15.2$ A.

$$\text{Power factor on full-load} = SP / OP = 6.45 / 7.6 = 0.86$$

$$\text{Max. torque / F.L. torque} = MK / PG = 10 / 5.6 = 1.8$$

$$\therefore \text{Max. torque} = 180\% \text{ of full-load torque.}$$

Example 8.3 Draw the circle diagram from no-load and short-circuit test of a 3-phase. 14.92k W, 400 V, 6-pole induction motor from the following test results (line values).

$$\text{No-load}: 400\text{-V}, 11\text{ A, p.f.} = 0.2$$

$$\text{Short-circuit}: 100\text{-V}, 25\text{ A, p.f.} = 0.4$$

Rotor Cu loss at standstill is half the total Cu loss. From the diagram, find

(a) line current, slip, efficiency and p.f. at full-load;
(b) the maximum torque.

SOLUTION:

$$\text{No-load p.f.} = 0.2; \theta_{nl} = \cos^{-1}(0.2) = 78.5^0$$

$$\text{Short-circuit p.f.} = 0.4; \theta_s = \cos^{-1}(0.4) = 66.4^0$$

$$\text{Short-circuit current with normal voltage, } I_{st} = 25 \times 400 / 100 = 100 \text{ A}$$

$$\text{Short-circuitpower input with this current} = \sqrt{3} \times 400 \times 100 \times 0.4 = 27{,}710 \text{ W}$$

Assume a current scale of 1 cm = 5 A. The circle diagram of above figure is constructed as follows:

(i) No-load current vector OO′ represents 11 A. Hence, it measures $11/5 = 2.2$ cm and is drawn at an angle of 78.5° with OY.
(ii) Vector OA represents 100 A and measures $100/5 = 20$ cm. It is drawn at an angle of 66.4° with OY.
(iii) O′D is drawn parallel to OX. NC is the right angle bisector of O′A.
(iv) With C as the center and CO′ as radius, a semicircle is drawn as shown.
(v) AF represents power input on short-circuit with normal voltage applied. It measures 8 cm and (as calculated above) represents 27,710 W. Hence, power scale becomes $1 \text{ cm} = 27{,}710/8 = 3{,}465$ W.

(a) FL motor output = 14,920 W. According to the above power scale, the intercept between the semicircle and the output line O′A should

measure $= 14{,}920/3{,}465 = 4.31$ cm. Hence, vertical line PL is found which measures 4.31 cm. Point P represents the full-load operating point.
Line current = OP = 6.5 cm which means that full-load line current $= 6.5 \times 5 = 32.5$ A.
$\theta = 32.9^{\circ}$ (by measurement)

$$\therefore \cos 32.9° = 0.84 \text{ (or } \cos\theta = PL / OP = 5.4 / 6.5 = 0.84)$$

$$\text{slip} = EK / PK = 0.3 / 5.35 = 0.056 \text{ or } 5.6\%;$$

$$\eta = PE / PL = 4.3 / 5.4 = 0.8 \text{ or } 80\%$$

(b) For finding maximum torque, line CM is drawn perpendicular to torque line O′H.

MT is the vertical intercept between the semicircle and the torque line and represents the maximum torque of the motor in synchronous watts.

Now, MT = 7.8 cm (by measurement)

$$\therefore T_{max} = 7.8 \times 3465 = 27{,}030 \text{ synch. W}$$

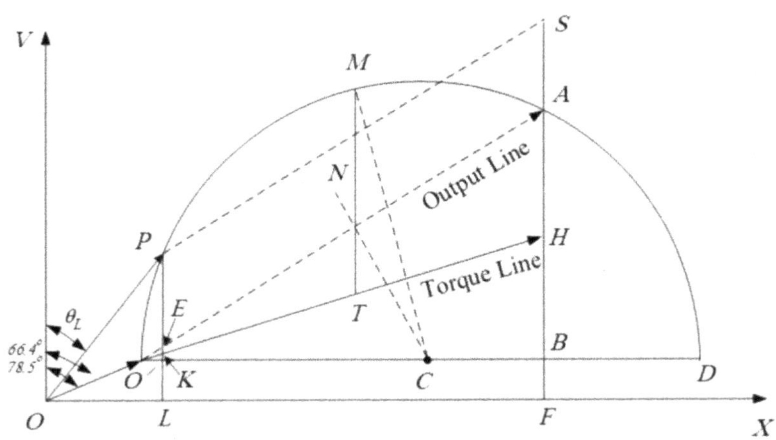

Example 8.4 A 415 V, 29.84 kW, 50 Hz, delta-connected motor gave the following test data:

No-load test: 415 V, 21 A, 1,250 W
Locked rotor test: 100 V, 45 A, 2,730 W
Construct the circle and determine

(a) the line current and power factor for rated output
(b) the maximum torque.

Assume stator and rotor Cu losses equal at standstill.

SOLUTION:

$$\text{Power factor on no-load is } \theta_{nl} = \cos^{-1}\frac{1250}{\sqrt{3}*415*21}$$

$$= \theta_{nl}\cos^{-1}(0.0918) = 84^0\ 44'$$

$$\text{Power factor with locked rotor is} = \frac{1250}{\sqrt{3}*100*45} = 0.3503$$

$$\theta_s = \cos^{-1}(0.3503) = 69^0\ 30'$$

The input current I_{st} on short-circuit if normal voltage were applied = 45 (415/100) = 186.75A and powe taken would be $= 2,730\,(415/100)^2 = 47,000$ W.

Let the current scale be $1\text{ cm} = 10\text{ A}$. The circle diagram of the figure below is constructed as follow:

(i) Vector OO′ represents 21 A so that it measures 2.1 cm and is laid at an angle of 84°44′ with OE (which is vertical, i.e., along Y-axis).

(ii) Vector OA measures $186.75/10 = 18.675\text{ cm}$ and is drawn at an angle of 69°30′ with OE.

(iii) O'D is drawn parallel to OX. NC is the right-angle bisector of O'A

(iv) With C as the center and CO′ as radius, a semi-circle is drawn as shown. This semi-circle is the locus of the current vector for all load conditions from no-load to short-circuit.

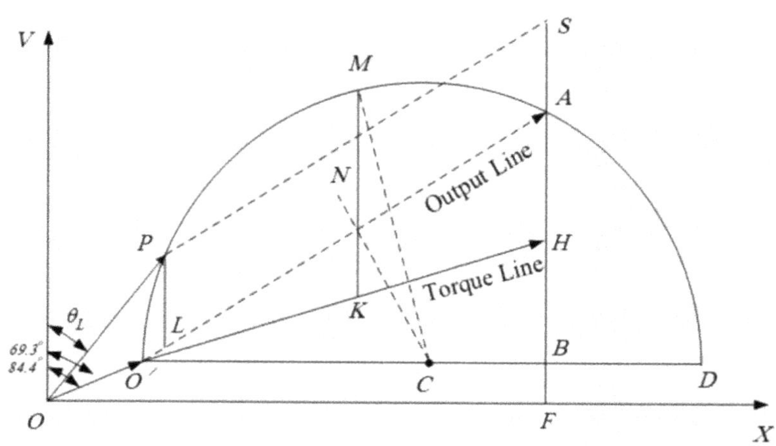

(a) The vertical AF represents power input on short-circuit with normal voltage applied.

(b) AF measures 6.6 cm and (as calculated above) represents 47,000 W. Hence, power scale becomes, $1\text{ cm} = 47,000/6.6 = 7,120\text{ W}$

(c) Full-load output $= 29{,}840$ W. According to the above power scale, the intercept between the semicircle and output line O′A should measure $29{,}840/7{,}120 = 4.19$ cm. Hence, line PL is found which measures 4.19 cm. Point P represents the full-load operating point.

Phase current $= OP = 6$ cm $= 6 \times 10 = 60$ A; Line current $= \sqrt{3}\sqrt{3} \times 60$
$= 104$ A; Power factor $= \cos\angle POE = \cos 35° = 0.819$

(d) For finding the maximum torque, line CM is drawn perpendicular to the torque line O'H. point H is such that

$$\frac{AH}{BH} = \frac{rotor\ CU\ losses}{stator\ CU\ losses}$$

Since the two Cu losses are equal, point H is the mid-point of AB. Line MK represents the maximum torque of the motor in synchronous watts.

$MK = 7.3$ cm (by measurement) $= 7.3 \times 7{,}120 = 51{,}980$ synch. Watt

Example 8.5 Draw the circle diagram for a 5.6 kW, 400 V, 3-phase, 4-pole, 50 Hz, slip-ring, induction motor from the following data:

No-load readings: 400 V, 6 A, $\cos\theta_{nl}\ \theta_{nl} = 0.087$.

Short-circuit test: 100 V, 12 A, 720 W.

The ratio of primary to secondary turns = 2.62, stator resistance per phase is 0.67Ω and of the rotor is 0.185Ω. Calculate

(i) full-load current; (ii) full-load slip; (iii) full-load power factor;
(iv) maximum torque/full – load torque; (v) maximum power

Solution:

$\theta_{nl} = \cos^{-1}(0.087) = 85°$

Short-circuit current with normal voltage $= 12 \times 400/100 = 48$ A

Total input $= 720 \times (48/12)^2 = 11.52$ kW

$$\cos\theta_s = \frac{720}{\sqrt{3} \times 100 \times 12} = 0.347 \text{ or } \theta_s = 69°\ 40'$$

Current scale is, 1 cm = 2 A.

In the circle diagram, OA = 3 cm and inclined at 85° with OV. Line OB represents short-circuit current with normal voltage. It measures 48/2 = 24 cm and represent 48 A. BD is perpendicular to
OX.

$$\text{For Drawing Torque Line } K = 2.62,\ R_1 = 0.67\ \Omega,\ R_2 = 0.185\Omega$$

(in practice, an allowance of 10% is made for skin effect)

$$\therefore \frac{rotor\ CU\ losses}{stator\ CU\ losses} = 2.62^2 \times \frac{0.185}{0.67} = 1.9$$

$$\therefore \frac{rotor\ CU\ losses}{total\ CU\ losses} = 0.655$$

$$\text{Now BD} = 8.25 \text{ cm and represents } 11.52 \text{ kW}$$

$$\text{power scale} = 11.52 / 8.25 = 1.4 \text{ kW / cm}$$

$$\therefore 1 \text{ cm} = 1.4 \text{ kW}$$

BE represents total Cu loss and is divided at point T in the ratio 1.9: 1.

$$BT = BE \times 1.9 / 2.9 = 0.655 \times 8 = 5.24 \text{ cm}$$

AT is the torque line Full-load output = 5.6 kW

It is represented by a line = 5.6/1.4 =4 cm

DB is produced to R such that BR = 4 cm. Line RP is parallel to output line and cuts the circle at

P. OP represents full-load current.

PS is drawn vertically. Points M and Y represent points of maximum torque and maximum output respectively.

(i) F.L. current = OP = 5.75 cm = $5.75 \times 2 = 11.5$ A
(ii) F.L. slip = FG/PG = 0.2/4.25 = 0.047 or 4.7%
(iii) p.f. = SP/OP = 4.6/5.75 = 0.8
(iv) $\dfrac{\max .torque}{full-load\ torque}$ = MK/PG = 10.05/4.25 = 2.37
(v) Maximum output is represented by YL = 7.75 cm.

$$\therefore \text{Max. output} = 7.75 \times 1.4 = 10.8 \text{ kW}$$

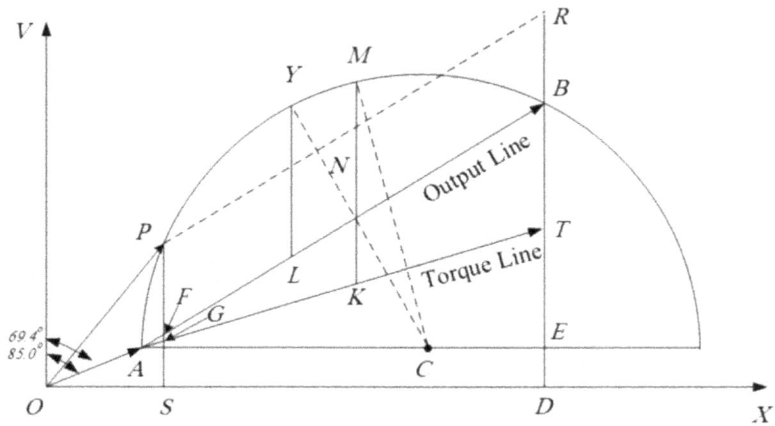

PROBLEMS

8.1 A 460 V 50 hp, six-pole, Δ-connected, 60 Hz, 3-phase, induction motor has a full-load slip of 4%, an efficiency of 91%, and a power factor of 0.87 lagging. At start-up, the motor develops 1.75 times the full-load torque but draws 7 times the rated current at the rated voltage. This motor is to be started with an autotransformer reduced voltage starter.

(a) What should the output voltage of the starter circuit be to reduce the starting torque until it equals the rated torque of the motor?
(b) What will the motor starting current, and the current drawn from the supply be at this voltage?

8.2 A wound-rotor induction motor is operating at rated voltage and frequency with its slip rings shorted and with a load of about 25% of the rated value for the machine. If the rotor resistance of this machine is doubled by inserting external resistors into the rotor circuit, explain what happens to the following:

(a) Slip s
(b) Motor speed N
(c) The induced voltage in the rotor
(d) The rotor current
(e) T_{ind}
(f) P_{out}
(g) P_{RCL}
(h) Overall efficiency η

9 Single-Phase Induction Motors

In the rural area, for example, there is no availability of 3-phase voltage supply. It is impossible to use 3-phase induction motors. For this reason, a motor with similar characteristics to the 3-phase motor was developed, but which works using single-phase supply.

For the single-phase motor to work, it needs special devices, such as capacitors, rings (clamp) in short circuit, auxiliary coils, and centrifugal switches, which allow it to rotate in the necessary direction. The presence of capacitors or other devices used in its startup fundamentally depends on the model and the work required.

9.1 SINGLE-PHASE INDUCTION MOTOR COMPONENTS

The single-phase motor consists of:

- stator, where the field coils are located that vary their arrangement, shape and connection according to the type of motor;
- squirrel-cage rotor, which is the moving part, consists of the armature.

Figure 9.1 shows the component parts of this type of motor.

9.2 WORKING PRINCIPLE OF SINGLE-PHASE INDUCTION MOTOR

Single-phase induction motors have a single winding on the stator. This winding generates a magnetic field that alternates with current alternations. In this case, the movement caused is not rotational. When the rotor is stationary, the magnetic field in the stator, expanding and contracting, induces currents in the rotor.

The field generated in the rotor is of opposite polarity to that of the stator. Thus, the opposition of the fields exerts a torque on the upper and lower parts of the rotor, which would tend to rotate it 180º from its original position. As the torque is equal in both directions, as the forces are exerted by the center of the rotor and in opposite directions, the rotor remains stationary. The schematic representation of this phenomenon in shown in Figure 9.2.

If the rotor is turning, it will continue to turn in the initial direction, as the torque will be helped by the inertia of the rotor and the induction of its magnetic field. As the rotor is rotating, the phase shift between the rotor and stator magnetic fields will not be greater than 180°.

DOI: 10.1201/ 9781003795841-9

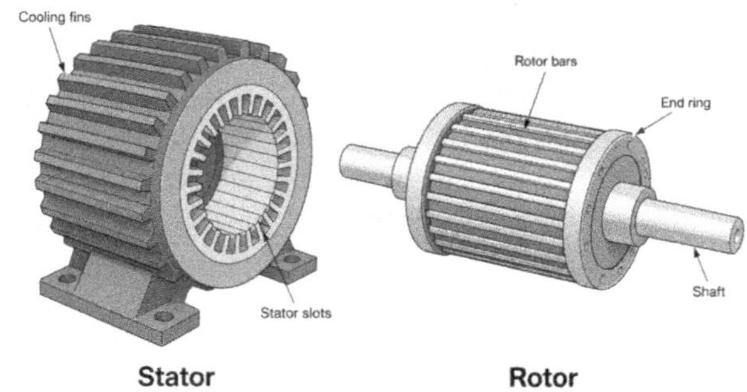

FIGURE 9.1 Stator and rotor of single-phase induction motor.

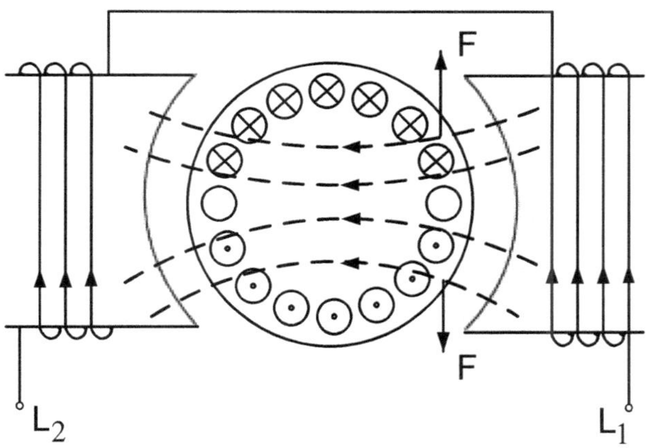

FIGURE 9.2 Rotor cut by the field in the single-phase motor.

To give the rotor the initial turn, two types of starters are commonly used:

- distorted field, which characterizes the distorted field engine;
- auxiliary phase with capacitor, which distinguishes the auxiliary phase motor.

The possibility of easily obtaining a rotating magnetic field is one of the main advantages of 3-phase current over single-phase. However, single-phase induction motors are manufactured despite technically being inferior to 3-phase motors. They are mainly used to drive small loads such as domestic fans, refrigerators, small water pumps, and so forth.

The stator of a single-phase motor is built to receive two windings: one, called the main winding, and another, called the auxiliary winding, which is essential for starting the motor, as will be seen below. The rotor is always caged. The magnetic field created by the main winding when traversed by a sinusoidal alternating current is a pulsating field whose magnetic axis is fixed in space, but whose direction varies according to the current. The flux from this field induces voltages and currents in the rotor bars, just as the magnetizing flux created by the primary winding of a transformer induces voltages and currents in the secondary winding.

Figure 9.3 shows the simplified schematic of a single-phase induction motor. The auxiliary winding is not indicated, only the two halves of the main winding. The alternating current that feeds the stator winding creates a pulsating MMF whose axis is fixed in space, varying its polarity, in time, according to the frequency of the current. In the rotor bars, the induced currents have such a direction that they produce an MMF contrary to the MMF of the stator. The axis of the MMF of the rotor coincides with the axis of the MMF of the stator, therefore, the load angle is equal to zero. There is, as a consequence, no torque that can act on the rotor and make it start. In this condition, the motor is simply a static transformer with the secondary (rotor) short-circuited.

This absence of starting torque in the single-phase motor can also be explained in the following way: the currents induced in the rotor create magnetic fields around the conductors that interact with the pulsating field of the rotor, giving rise to forces that act on these conductors. In the case of Figure 9.3, considering the magnetic axis of the stator MMF as a vertical one, the currents induced in the rotor conductors to the left of the vertical will have the opposite direction to the currents induced in the conductors on the right. Therefore, the forces that act on them are equal and opposite in direction and their resultant is zero. The absence of a starting torque is a characteristic of single-phase induction motors. When connected directly to the mains, the motor will not start, unless it is given an initial impulse by an external force that will make it rotate in the direction of the applied force.

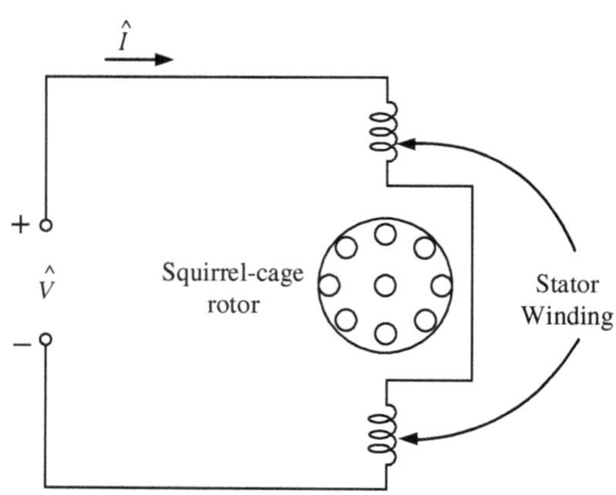

FIGURE 9.3 Simplified diagram of a single-phase induction motor.

In addition to the above explanations, the absence of starting torque in a single-phase motor can be understood in light of the theory of the two rotating magnetic fields present in the operation of a single-phase induction motor. This theory is extremely useful to not only explain not only the absence of the starting torque but also serve as a basis for elaborating the equivalent circuit of the motor.

By this theory, the pulsating MMF created in the stator by the alternating current $i = I_m \sin\omega t$ is composed of two MMFs, equal in module, which rotate, in space, in synchrony with the network frequency, but in opposite directions.

As seen when studying the rotating field of the 3-phase induction motor, the MMF expression of a pulsating magnetic field, created by a sinusoidal alternating current, is given by:

$$F_1 = F_{1\,max} \cos\omega t \cos\theta \tag{9.1}$$

$F_{1\,max}$ represents the maximum value of the MMF and θ the spatial electric angle measured from the magnetic axis of the stator coil. According to a well-known trigonometric relationship, equation (9.2) is equal to:

$$F_1 = \frac{F_{1max}}{2}\cos(\theta - \omega t) + \frac{F_{1max}}{2}\cos(\theta + \omega t) \tag{9.2}$$

The second part of equation (9.2) are the expressions of the MMFs of two rotating magnetic fields whose amplitudes are equal to half the amplitude of the MMF of the pulsating magnetic field and whose directions of rotation are opposite. Figure 9.4 shows the phasor sum, at four different instants of the fluxes of the two rotating fields that form the pulsating flux.

Each of the fluxes $\varphi 1$ and $\varphi 2$ is half of the pulsating flux φ. This sum is true only for the stopped rotor. With the rotor stopped, both rotating fluxes will induce equal currents in the bars, but in opposite directions, which, interacting with the pulsating field, will produce equal torques in opposite directions that do not allow the rotor to start. If an external force acts on its axis, clockwise or counterclockwise, the rotor will continue to rotate in the direction of the applied force, by itself, until it reaches a speed close to synchronism. This flow that rotates in the same direction as the rotor is called forward rotating flow and the other that rotates in the opposite direction, reverse rotating flow.

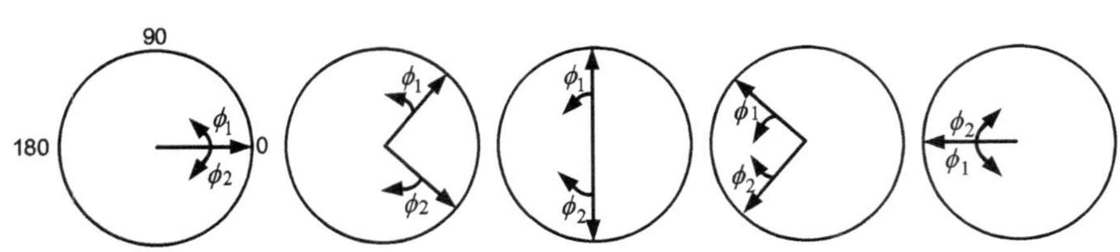

FIGURE 9.4 Opposite rotating flows whose sum is equal to the pulsating flow.

As the rotor rotates, the reverse flux will be heavily demagnetized and the resulting flux will be virtually forward rotating flux, for speeds close to synchronous. The demagnetization of the reverse flux is explained as follows: the slip of the rotor, rotating at speed n in the same direction as the forward flux, with respect to the synchronous speed n1 of the reverse flux, is equal to:

$$s = \frac{-n_1 - n}{-n_1} = \frac{n_1 + n}{n_1} = \frac{n_1 + n_1(1-s)}{n_1} = 2 - s \tag{9.3}$$

Therefore, the slip frequency produced by reverse flow is much higher than that produced by forward flow. For example, for a mains frequency of 60 Hz and a slip of 0.05 produced by forward flow, the slip frequency relative to forward flow would be $0.05 \times 60 = 3$ Hz, while that relative to reverse flow would be $(2-0,05)60 = 117$ Hz, that is, 39 times greater than that of the direct flow. At this frequency, the rotor reactance is much greater than its resistance and, as a consequence, the currents due to this field will be practically reactive, exerting a strong demagnetizing effect on the reverse flux. Thus, for small values of slip, the torque of a single-phase induction motor is produced, practically, by direct flux. There is also a small reverse flow that produces a braking effect, as the torque it produces opposes the torque produced by the forward flow.

Figure 9.5 shows in dashed lines the characteristic curves of the torques produced by forward and reverse flows (forwards and backwards, in the figure) and in solid

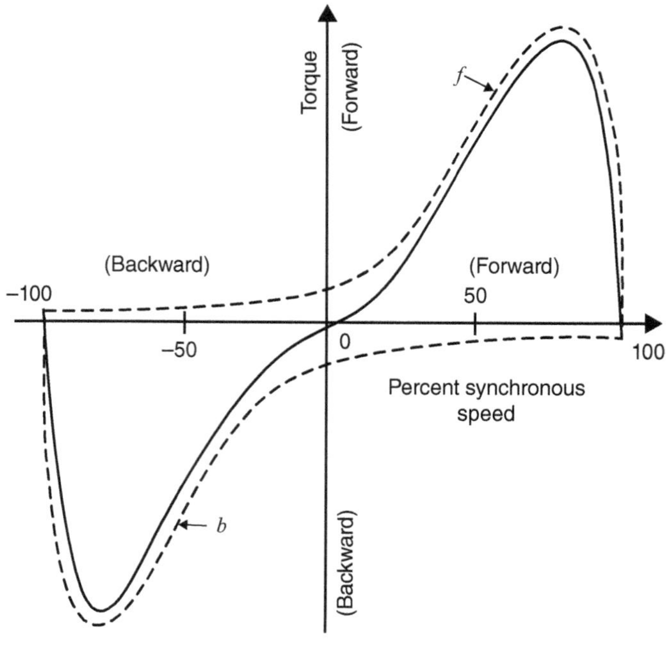

FIGURE 9.5 Direct, reverse and resultant conjugate curves

lines the resulting characteristic. As can be seen at synchronous speed ($n = n1$), unlike what happens in the 3-phase motor, there is a braking torque due to the reverse torque acting on the rotor of the single-phase motor.

The current circulating in the rotor is formed by the superposition of two currents of very different frequencies. The rotor power loss is the sum of the currents induced by the two fields separately. For this reason, the power loss in the rotor of a single-phase motor is much greater than that which occurs in a 3-phase motor, which makes it a lower efficiency motor than the corresponding 3-phase motor.

An increase in the slip of a single-phase motor, in addition to increasing the current I1, increases the braking torque of the inverse field. For this reason, the operation of a single-phase motor is less stable, and its maximum torque is lower than that of a 3-phase motor.

The starting of a single-phase motor is carried out with the help of an auxiliary winding. The role of this auxiliary winding is to create a flux that is out of phase in space and time with the main flux and thus obtain a resulting rotating flux at start-up. For this purpose, the auxiliary winding is mounted in the stator slots in such a way as to create a magnetic flux that is approximately 90° electrically out of phase with the main flux. However, to create the resulting rotating flux, it is not enough for this flux to be 90° electrically out of phase in space. It is necessary that it be traversed by a current that is as much as possible out of phase 90 electrically, in time, from the current flowing through the main winding. The resulting field produces a torque of small value, but sufficient to start the motor in the forward direction. As the motor accelerates, the torque due to the forward field increases and that due to the reverse field decreases. The motor thus reaches its normal speed. The auxiliary winding is automatically turned off shortly before the motor reaches its normal speed, as its actuation is no longer required. For this reason, it is designed to stay on only during the time the engine takes to accelerate.

9.3 AUXILIARY PHASE SINGLE-PHASE MOTORS

The auxiliary phase single-phase motor is the most widely used. Its mechanical construction is the same as that of 3-phase induction motors. Thus, in the stator, there are two windings: one with thicker wire and with a large number of turns (main or working winding), and another with thinner wire and with few turns (auxiliary or starting winding).The main winding is switched on during the entire motor running time, but the auxiliary winding only operates during starting. This winding is turned off when an automatic device is activated, located partly on the engine cover and partly on the rotor. Generally, a capacitor is connected in series with the auxiliary winding, thereby improving the motor starting torque. There are basically four types of single-phase squirrel-cage induction motors (auxiliary phase), classified according to the auxiliary starting arrangement used:

- divided phase motor (split-phase);
- motor with starting capacitor (capacitor-start);
- motor with permanent capacitor (permanent-split);
- motor with two capacitors (two-value capacitor).

9.4 DISTORTED FIELD MOTOR OR SHADED POLE

The distorted field motor consists of a squirrel-cage rotor and a stator similar to the universal motor. However, in the distorted-field motor, there is a groove in the pole shoe in which a short-circuited copper ring or loop is housed. Therefore, this motor is also known as a ring motor or a short-circuited loop motor.

Since, in the distorted-field motor, the rotor is of the squirrel-cage type, all the connections are in the stator. See the schematic representation of these connections in (Figure 9.7).

The distorted field motor stands out among single-phase induction motors for its simple, reliable and economical starting process, without capacitors or starting switches. Constructively, its most common form is that of salient poles, schematically illustrated in (Figure 9.8).

Note that a portion of each pole (typically 25% to 35%) is embraced by a short-circuited copper loop, called a short-circuit or shading clamp. The current induced in this loop causes the flux passing through it to suffer a delay in relation to the flux in the part not embraced by the loop. The result of this is similar to a rotating field

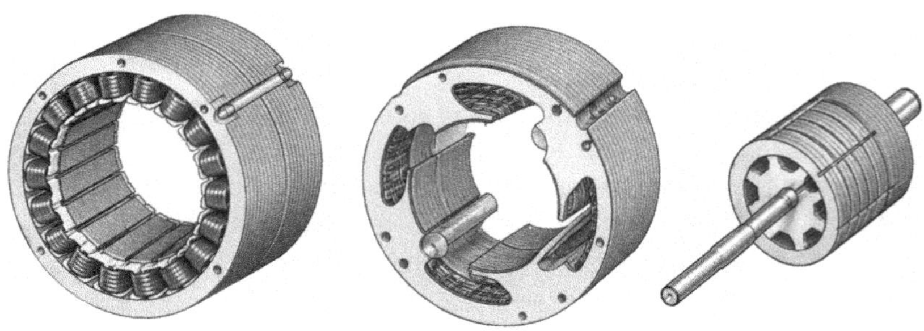

FIGURE 9.6 Stator with windings: stator and rotor of the distorted field motor

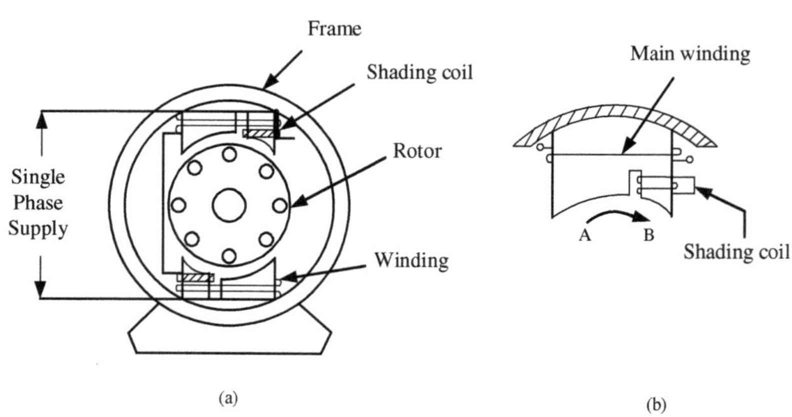

FIGURE 9.7 Distorted field motor.

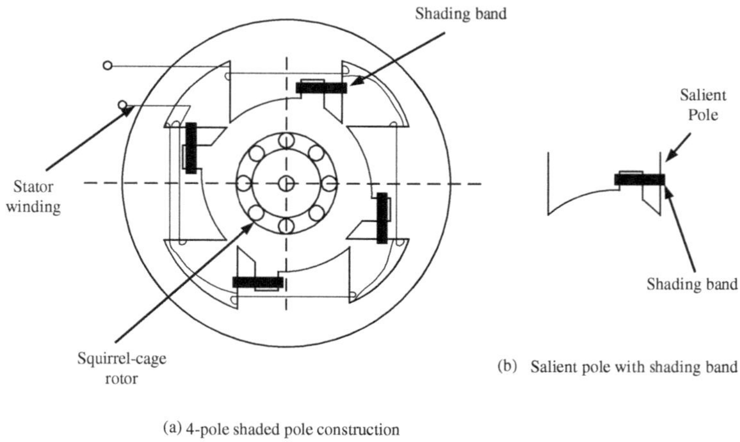

FIGURE 9.8 Salient pole motor.

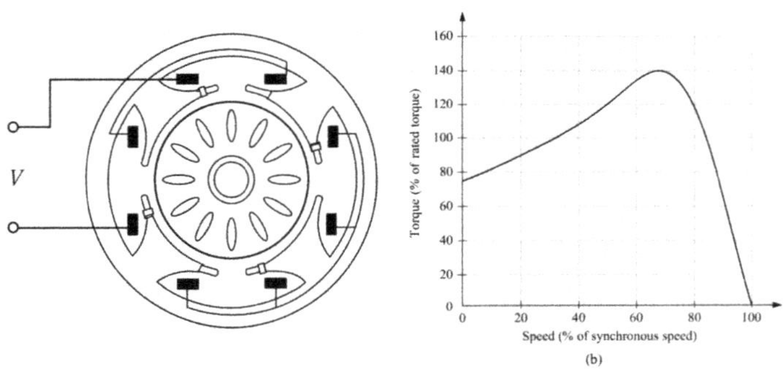

FIGURE 9.9 Torque – speed characteristic.

moving from the non-embraced to the embraced part of the pole, producing torque that will cause the motor to start and reach rated speed. Because of this, the distorted field motor has only one direction of rotation. As for performance, distorted field motors have low starting torque (15% to 50% of nominal), low efficiency (10% to 30%), and low power factor (25% to 50%). That is why they are manufactured only for small powers, ranging from a few thousandths of hp (horsepower) to a maximum limit of 1/4 hp.

9.5 SPLIT PHASE MOTOR (SPLIT-PHASE)

This motor has a main winding and an auxiliary (for starting) winding both are electrically 90° out of phase in space. The auxiliary winding creates a phase

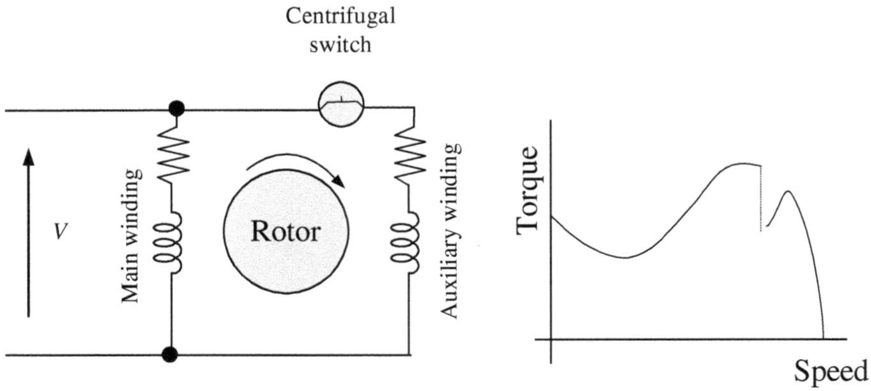

FIGURE 9.10 Basic diagram, split-phase motor torque versus speed characteristic.

shift that produces the torque needed for initial rotation and acceleration. When the motor reaches a predetermined speed, the auxiliary winding is disconnected from the mains using a switch that is normally actuated by centrifugal force (switch or centrifugal circuit breaker) or, in specific cases, by current relay, electronic switch, manual switch or other special devices. Figure 9.10 shows the schematic representation of the basic wiring diagram for this motor with the torque versus speed ratio curve.

As the auxiliary winding is sized to operate only at start-up, not turning it off will cause it to burn out. The phase shift angle that can be obtained between the main winding and auxiliary winding currents is small and, therefore, these motors have a starting torque lower or slightly higher than the nominal one, which limits their application to powers of up to 1/3 hp and loads that require reduced or moderate starting torque.

9.6 MOTOR WITH STARTING CAPACITOR (CAPACITOR-START)

It is a similar motor to the split phase. The main difference lies in the inclusion of an (electrolytic) capacitor in series with the starting auxiliary winding. The capacitor allows for a greater phase shift between the main and starting winding currents and, therefore, a high starting torque.

As with the split-phase motor, the auxiliary circuit is disconnected when the motor reaches approximately 80% of its service speed. After that, its operation is similar to that of a split-phase motor as shown in Figure 9.11.

Due to its high starting torque (between 200% and 350% of the nominal torque), the capacitor starting motor can be used in a wide variety of applications, and is manufactured in powers ranging from 1/4 hp to 15 hp.

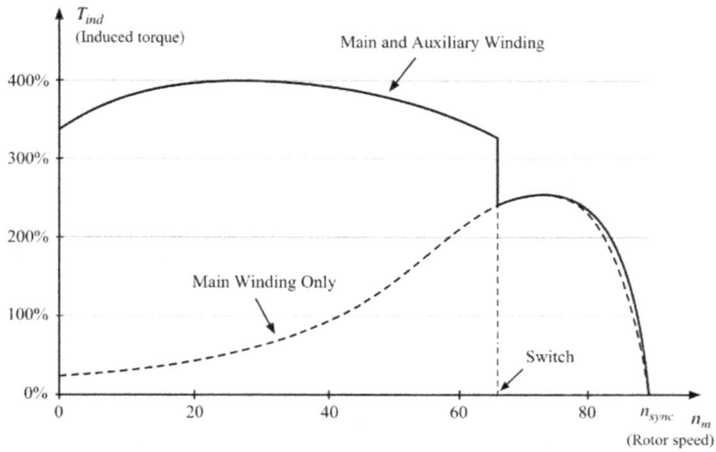

FIGURE 9.11 Basic diagram, motor torque – speed characteristic of capacitor start motor.

9.7 PERMANENT CAPACITOR MOTOR (PERMANENT-SPLIT)

The permanent capacitor motor is similar to the starting capacitor motor, with the difference that the auxiliary circuit is permanently connected to the electrostatic capacitor used. Observe in Figure 9.12 is the schematic representation of the motor with the capacitor and the torque × speed curve.

The operation of the permanent capacitor motor is similar to that of a 2-phase motor (it has a rotating field and not a pulsating one), which gives it silent operation, high efficiency, high power factor, and low starting current. From a constructive point of view, this motor is smaller than the other auxiliary phase motors and maintenance-free – as it does not use contacts and moving parts, as the motors previously studied. Normally, the permanent capacitor motor is manufactured for powers from 1/50 hp (read one- fifty horsepower) to 1.5 hp.

9.8 MOTOR WITH TWO CAPACITORS (TWO-VALUE CAPACITOR)

The two-capacitor motor uses the advantages of the two previous ones: starting like the starting capacitor motor and steady-state operation like the permanent capacitor motor. However, due to its high cost, it is only manufactured in powers greater than 2 hp. The following illustration shows the basic schematic of this motor and the torque-speed ratio curve.

Based on the characteristics of the asynchronous single-phase motor, analyze its advantages and disadvantages.

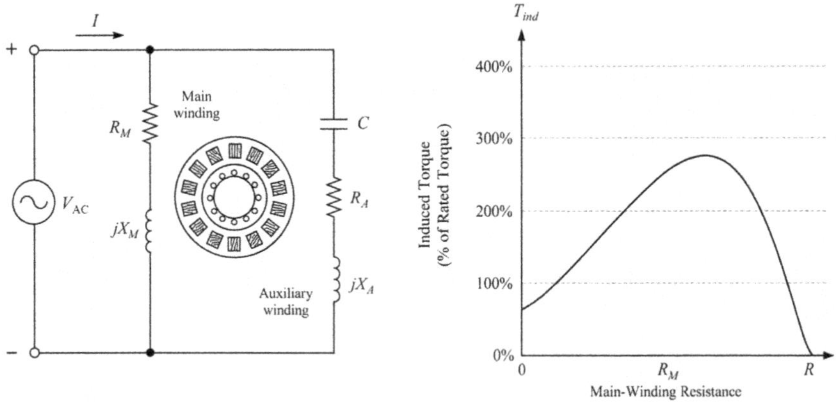

FIGURE 9.12 Basic diagram, torque – motor speed with permanent capacitor motor.

9.9 EQUIVALENT CIRCUIT OF SINGLE-PHASE INDUCTION MOTOR

The analysis presented in this section assumes that the single-phase induction motor is operating with only the main winding powered. With the rotor stopped and only the main winding powered, the motor is equivalent to a transformer with the secondary short-circuited, where there is an air gap in the magnetic circuit.

According to the previously presented double rotating field theory, it was shown that the MMF, stator pulse can be considered equivalent to two rotating MMFs of constant amplitude and equal to half of the first, which rotate in opposite directions.

With the rotor stopped, the amplitudes of the resulting forward and retrograde flows are equal to half the amplitude of the pulsating flow. The counter EMF resultant is created in the main winding by the pulsating flux in the air gap due to the stator and rotor MMFs. This is equal to the sum of two components, created by the forward and retrograde flows, that is, $E_1 = E_d + E_r$. In this situation with the rotor stopped, the value of the counter EMFs associated with the forward and retrograde fluxes are the same, $E_d = E_r$, as can be seen in Figure 9.14, which represents the equivalent circuit of a single-phase induction motor with the rotor stopped.

In this circuit, r_1 and x_1 represent, respectively, the leakage resistance and reactance of the main winding, x_m represents the magnetization reactance, r_{2e} and x_{2e} represent, respectively, the rotor leakage resistance and reactance, referred to the stator, with the rotor stopped. The core is not accounted for here and may be included later in rotational losses. Assuming that the rotor was turned by applying an external torque, slip is. The currents induced in the rotor by the direct field have frequencies f and sf where f is the frequency of the stator supply.

As in the case of the 3-phase induction motor, these rotor currents produce an MMF. rotating with speed proportional to sf in relation to the rotor, being synchronous

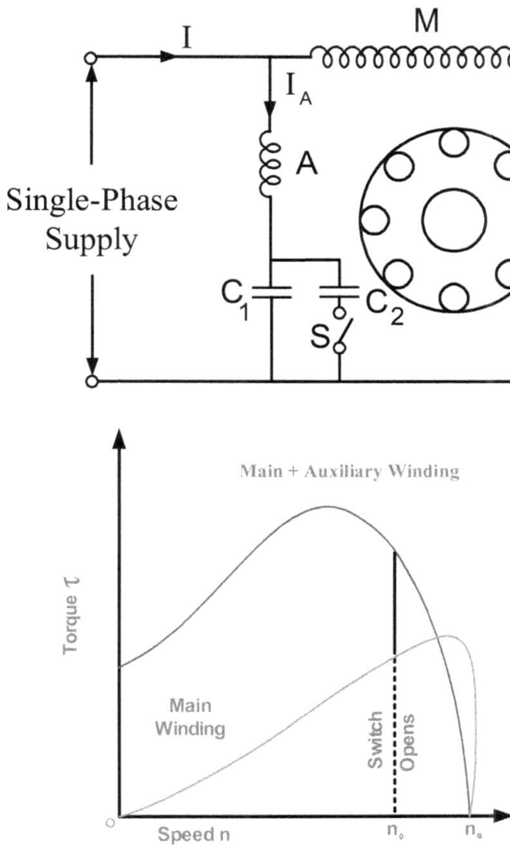

FIGURE 9.13 Basic scheme, torque versus speed characteristic of the motor with two capacitors.

TABLE 9.1
Advantages and disadvantages of the asynchronous single-phase motor

Single-phase induction motor	
Disadvantages	**Benefits**
• High starting current.	• Low acquisition cost.
• Higher maintenance cost due to the need for starting devices.	• Good starting torque.
• Imbalance in electrical installations (since it is single-phase).	• Constant speed.
• Low power factor when not well dimensioned for the load.	

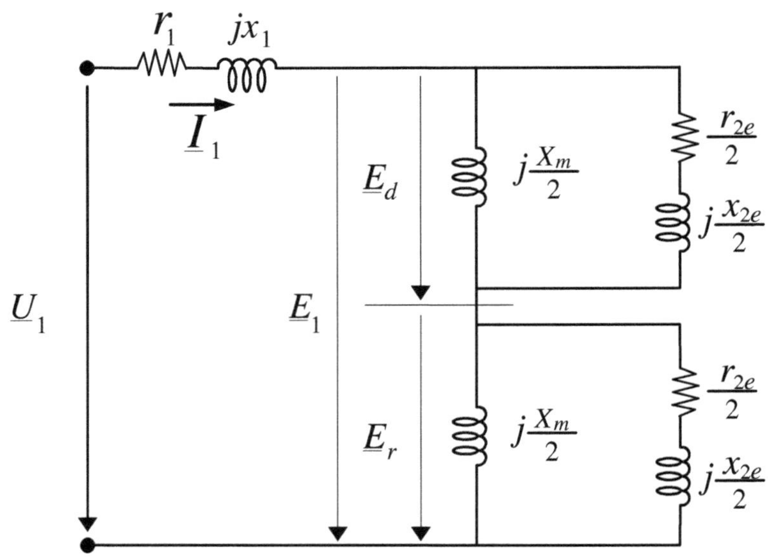

FIGURE 9.14 Equivalent circuit of the single-phase induction motor with the rotor stopped.

with that of the stator. The MMF resultant of these two creates a resultant flux rotating in the forward direction that gives rise to the EMF in the main stator winding. The effect created on the rotor seen from the stator can be represented by the impedance $r_2e/2 + x_2e/2_s$, as shown in (Figure 9.15).

When the rotor is rotating with slip s with respect to the forward field, its slip with respect to the retrograde field is 2-s. Thus, the retrograde field induces currents in the rotor whose frequency is $(2\text{-s})f$. For small slips, these rotor currents are almost twice the frequency of the supply. Seen from the stator, the MMF of the rotor caused by the retrograde field rotates at the synchronous speed, but in the opposite direction of rotation of the rotor. As mentioned in relation to the direct field, the equivalent circuit seen from the stator that represents these reactions is the equivalent of a polyphase motor with 2-s slip. In this circuit, the losses are already represented. magnetic with the inclusion of resistance. It is also shown, E_r which represents the EMF generated in the main stator winding due to the resultant retrograde field.

Analyzing the equivalent circuit, it is verified that, when the rotor rotates with small slips, the effect of the rotor resistance in the direct field, $r_{2e}/2s$, is much higher compared to the situation with the rotor stopped, while the effect in the retrograde field, $r_{2e}/2\,(2\text{-s})$, is lower compared to the locked rotor situation. In this way, the impedance associated with the forward field is higher in relation to the stationary rotor situation, while the impedance associated with the retrograde field, Zr, is lower.

Thus, the forward counter-electromotive force (forward cEMF). Ed is higher when the rotor is moving, while against forward counter-electromotive force, it is inferior. This means that the flux associated with the forward field increases, as opposed to

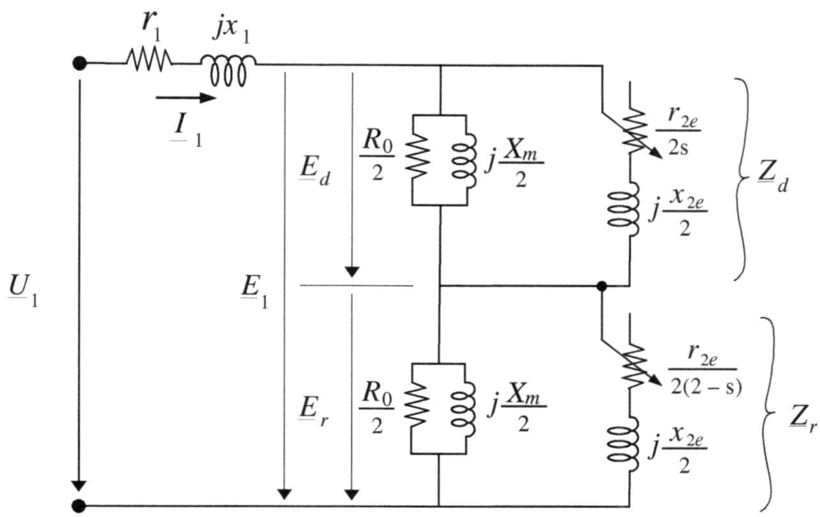

FIGURE 9.15 Equivalent circuit of single-phase induction motor.

that associated with the retrograde field, which decreases when the rotor accelerates, as previously illustrated in Figure 9.5.

The mechanical power and torque can be obtained using the same relationships studied for polyphase machines. Thus, the powers in the air gap for the forward and backward fields can be obtained by the following expressions:

$$P_d = I_1^2 R_d$$

$$P_r = I_1^2 R_r$$

where R_d and R_r correspond to the real part of the impedances associated with the forward Z_d and retrograde fields, Z_r respectively.

The torques produced by the two fields can be obtained as follows:

$$T_d = \frac{1}{\omega_s} P_d$$

$$T_r = \frac{1}{\omega_s} P_r$$

where ω_s is the synchronous speed in rad/s. As the two torques oppose each other, the resulting torque will be given by:

$$T_{el} = T_d - T_r = \frac{1}{\omega_s}\left(P_d - P_r\right) = \frac{I_1^2}{\omega_s}\left(R_d - R_r\right)$$

Since the rotor currents due to the forward and retrograde fields have different frequencies, the power losses in the rotor will be equal to the sum of the losses due to each of the fields. Then, the power losses in the rotor due to the direct field are:

$$p^{d}_{J_{rotor}} = sP_{d}$$

while the power losses in the rotor due to the retrograde field are:

$$p^{r}_{J_{rotor}} = (2-s)\,P_{r}$$

In this way, the total power losses in the rotor are equal to:

$$p_{J_{rotor}} = sP_{d} + (2-s)P_{r}$$

Since the angular velocity of the rotor is equal to (1-s) ω_s, the mechanical power developed by the motor can be calculated by the following expression:

Since the angular velocity of the rotor is equal to (1-s) ω_s, the mechanical power developed by the motor can be calculated by the following expression:

$$P_{el} = (1-s)\omega_{s}T_{el} = (1-s)\left(P_{d} - P_{r}\right)$$

As with the 3-phase induction motor, the mechanical power developed is different from the useful power due to the existence of rotational losses. Like this:

$$P_{u} = P_{el} - p_{rot}$$

10 Stepper Motors

10.1 INTRODUCTION

Stepper motors are a type of electric motor that converts electrical pulses into rotational motion. They are known for their precise and accurate control of movement, as they move in small, precise steps or increments, rather than continuously rotating like other types of motors.

Stepper motors have a number of advantages over other types of motors, including their ability to hold their position without the need for a feedback mechanism, and their ability to accurately control speed and position. They are commonly used in a wide range of applications, including robotics, CNC machines, 3D printers, and other types of precision machinery.

One of the key factors to consider when choosing a stepper motor is its step angle, which refers to the angle of rotation of the motor for each electrical pulse applied. A smaller step angle will result in more precise movement, while a larger step angle will result in faster movement.

Another important consideration is the torque of the motor, which refers to its ability to generate rotational force. Higher torque motors are able to move heavier loads or operate at higher speeds. Overall, stepper motors are an excellent choice for applications that require precise control over movement, and their versatility and ease of use make them a popular choice for a wide range of industries and applications.

10.1.1 Operation of Stepper Motors

Inside, there are stators formed by coils that, when an electric current passes through them, generate the necessary field for the movement of the rotor. This rotation is controlled by means of an external circuit that promotes the oscillation of the signal that will travel through the stator pairs, which is why they cannot be connected directly to the power supply, so there would not be the necessary pulsation for the rotor to rotate. Figure 10.1 demonstrates a rotation of one step in the motor through the oscillation of the signal applied to the stator.

The movement occurs when there is a cycle in the field of the stators, that is, each time a pair of stators is magnetized, an axis step occurs in the motor. It is necessary to build an oscillator for the stepper motor, according to the number of steps you want to acquire. For example, if we are going to use a stepper motor with six stators on an eight-pole rotor, then we are working with a motor that takes 48 steps per rotation, that is, a step at 7.5°.

10.1.2 Full-Step and Half-Step

Energizing one and only one coil at a time produces a small displacement in the rotor. This displacement occurs simply because the rotor is magnetically active, and

DOI: 10.1201/9781003795841-10

the energizing of the coils creates an intense magnetic field that acts in the sense of aligning itself with the teeth of the rotor. Thus, properly polarizing the coils, we can move the rotor between the coils (half-step) or aligned with them (full-step). Below are the movements performed.

Full Step: This drive mode is characterized by the fact that the motor moves its rotor in full drive step that it receives in its phases from the drive circuit.

Half-Step: Consists of alternately energizing one and two phases, thus allowing to move the rotor in half a step at a time. In this drive mode, as you can easily see, the number of "steps" is doubled to complete a turn.

When running at full pitch, the pitch resolution and torque are lower, it also consumes less power. Figures 10.2 and 10.3 demonstrate the operation of unipolar and bipolar full-step motor.

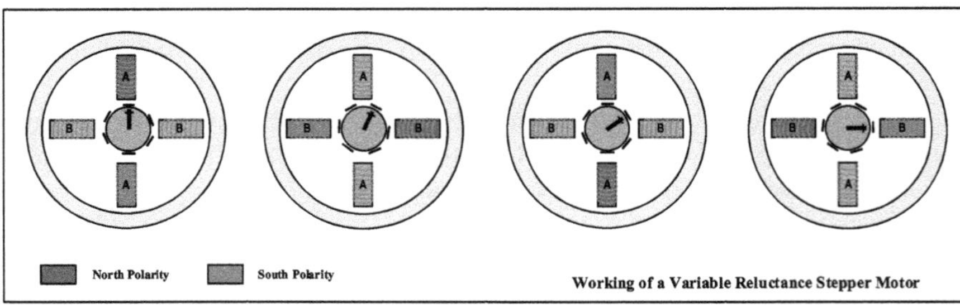

FIGURE 10.1 Stepper motor operation.

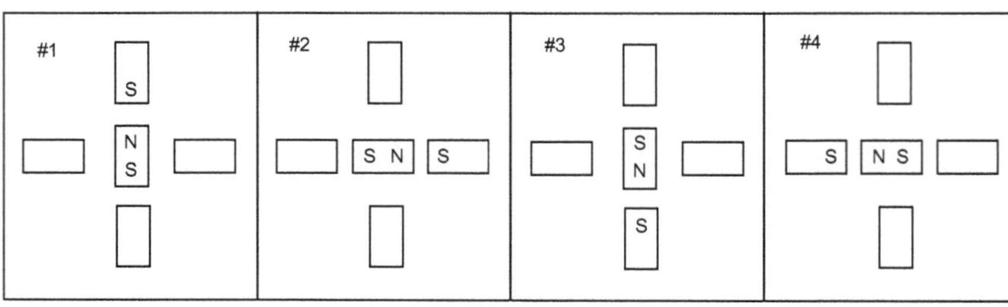

FIGURE 10.2 Unipolar Full Step.

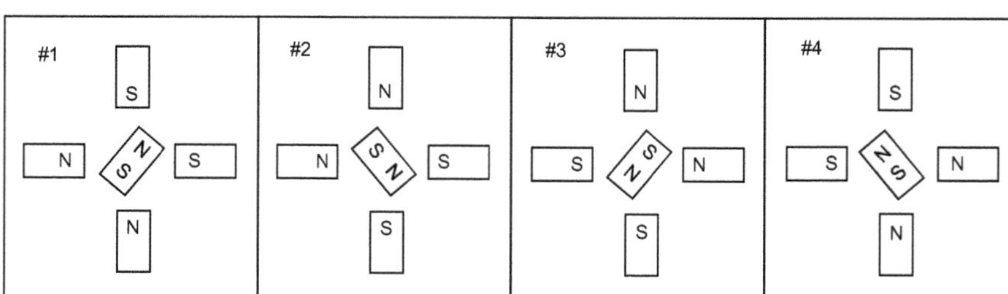

FIGURE 10.3 Bipolar full-step motor.

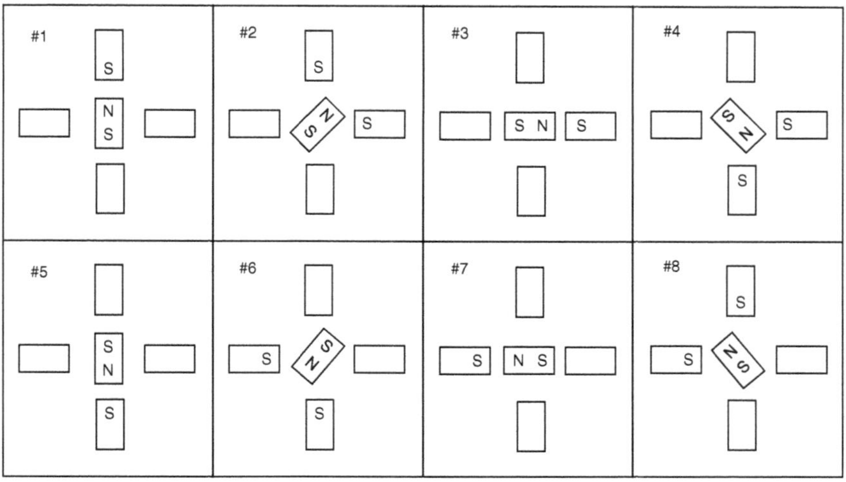

FIGURE 10.4 Unipolar half-step motor.

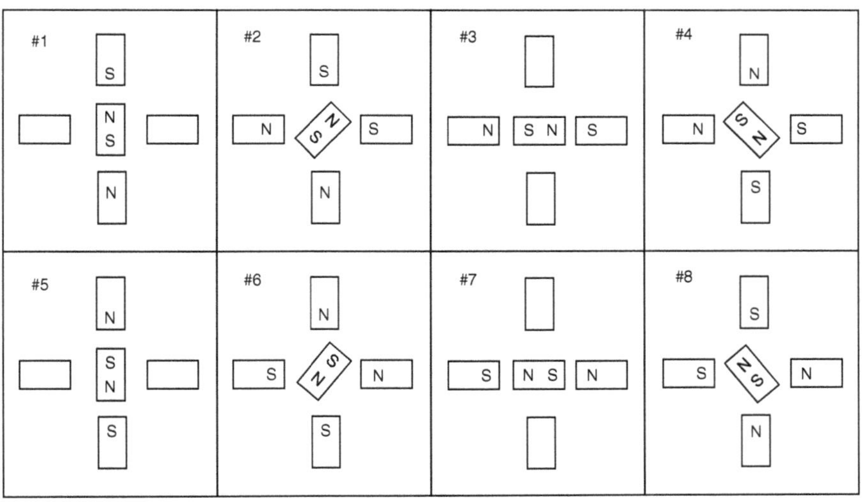

FIGURE 10.5 Bipolar half-step motor.

When it works with half a step, the step resolution is greater and the torque increases, on the other hand, there is more energy consumption, as there is a greater activation of the coils. Control for a half-step move also becomes more complex. Figures 10.4 and 10.5 demonstrate the operation of unipolar and bipolar half-step motor.

10.1.3 Unipolar Stepper Motor

Unipolar stepper motors are easily recognized by the center tap on each of the coils. The number of phases is twice as large as the number of coils, since each coil is divided into two. In Figure 10.6, we have the representation of a unipolar stepper motor with four phases (1a, 2a, 1b, and 2b). Phase 1a runs from center tap to end a on coil 1, and phase 1b runs from center tap to end b on this same coil. The phases of coil 2 are analogous to coil 1.

10.1.4 Bipolar Stepper Motor

Bipolar stepper motors are center-tapped coil motors. For this reason, these coils must be energized in such a way that the electric current flows in the opposite direction every two steps to allow continuous movement of the motor, that is, the polarity must be reversed during motor operation.

The bipolar example presented has two coils, as shown in Figure 10.7. In the case of the bipolar motor, now we have the number of phases equal to the number of coils that make up the motor winding, Phase 1ab and 2ab.

Bipolar stepper motors are known for their excellent size/torque ratio; they provide a higher torque, about 40% more compared to a unipolar motor of the same size. This is due to the fact that when energizing a phase, both poles where the phase (or

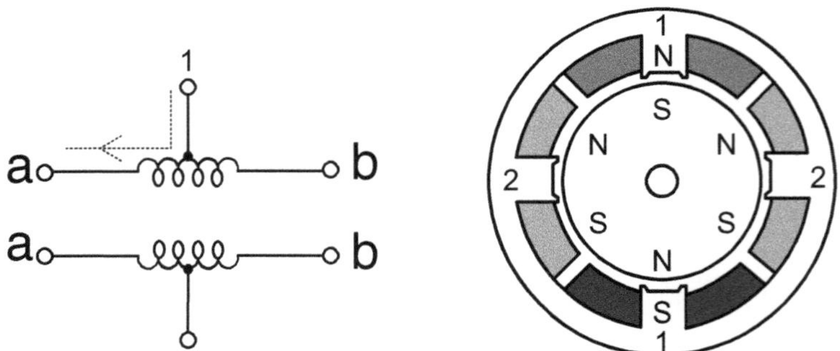

FIGURE 10.6 Four-phase unipolar stepper motor.

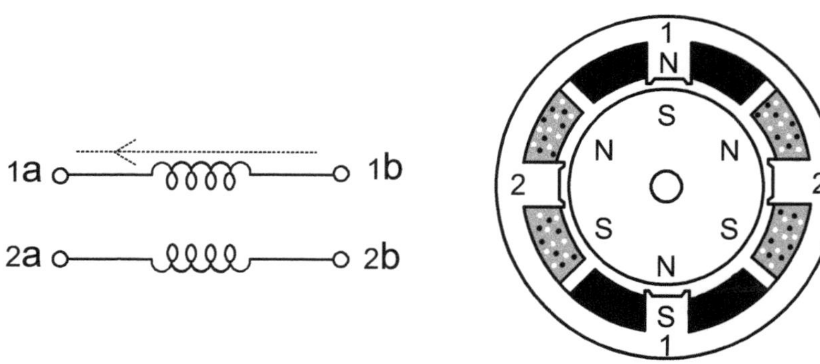

FIGURE 10.7 Two-phase bipolar stepper motor.

coil) is installed are magnetized. Thus, the rotor undergoes magnetic action from both poles, instead of just one, as in a unipolar motor.

10.2 PERMANENT MAGNET STEPPER MOTORS

These stepper motors are characterized by their rotor being formed by a permanent magnet. The stator is formed by teeth wrapped by coils. You can see the basic structure of a four-phase stepper motor in Figure 10.8.

When energizing one of the phases, a magnetic field is created. Thus, a torque is generated by the attraction of the generated pole with the opposite pole of the permanent magnet. A basic driver to control this stepper motor is shown in Figure 10.9.

If switch "S1" is turned on, and then switch "S2" is turned on, the motor will perform a 45° rotation. Figure 10.10 illustrates how this is done.

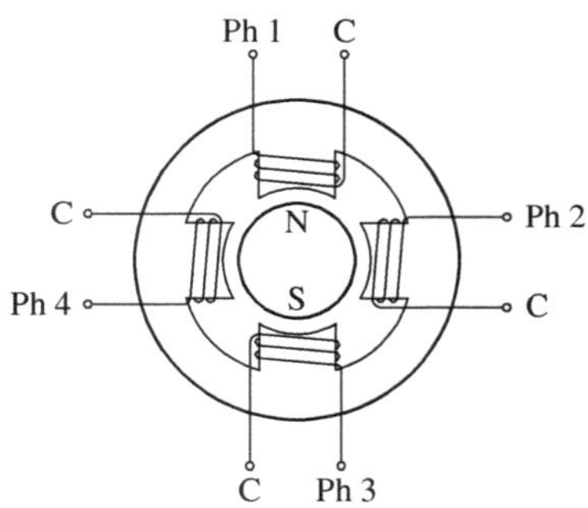

FIGURE 10.8 Four-phase motor.

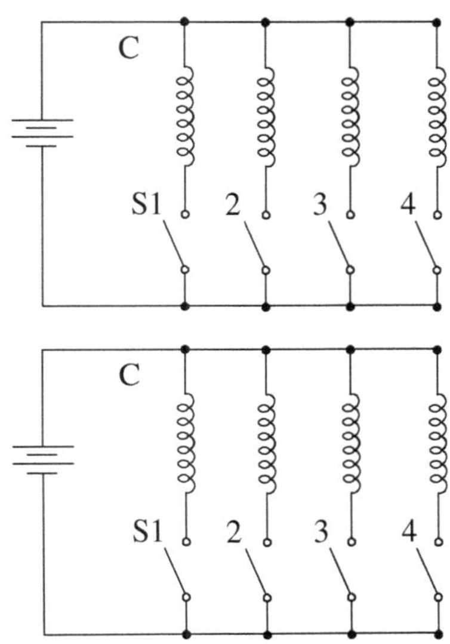

FIGURE 10.9 Four-phase driver.

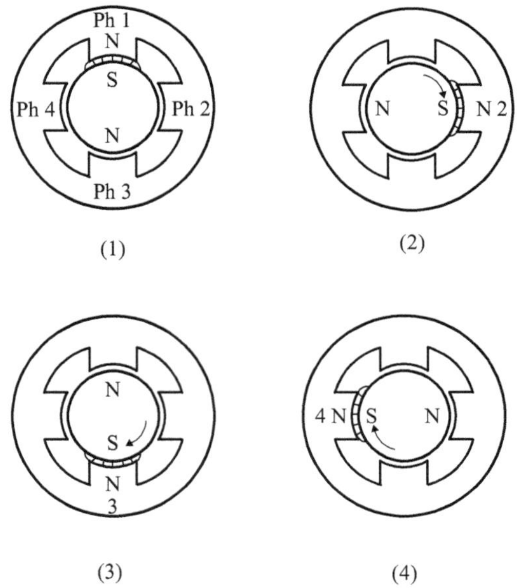

FIGURE 10.10 Sequence 4-phase motor.

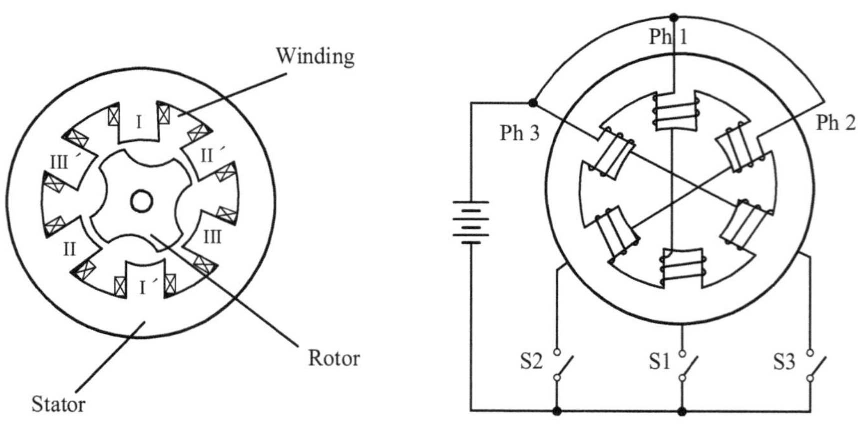

FIGURE 10.11 Variable reluctance motor.

10.3 VARIABLE RELUCTANCE STEPPER MOTORS

The simplest stepper motors are variable reluctance (VR) motors. Figure 10.11 shows the schematic section of such a motor with six stator teeth.

In Figure 10.11, each pair of opposing stator teeth are connected in the same phase. The rotor has four teeth. The stator and rotor core are usually made of steel.

FIGURE 10.12 Variable reluctance motor flux.

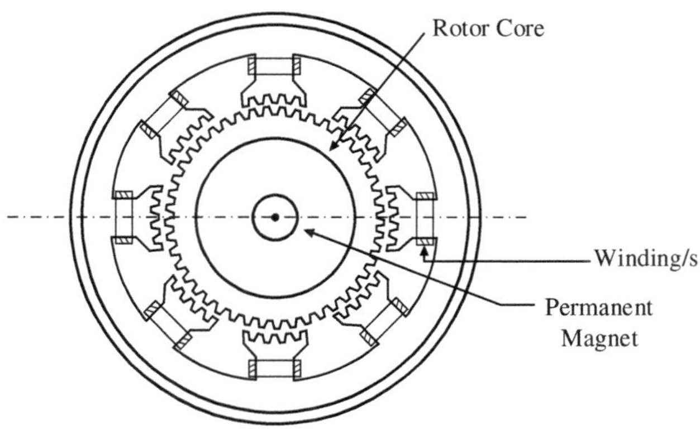

FIGURE 10.13 Hybrid motor.

If current is applied to phase 1, Ph1 in Figure 10.11, a magnetic field is created and a rotor tooth tries to align with it. Figure 10.12 shows how this happens.

When the rotor aligns with the magnetic field, the magnetic reluctance is minimized. If the rotor becomes misaligned with the magnetic field, a torque is generated to return it to the equilibrium position.

10.4 HYBRID MOTOR

This type of motor mixes the most sophisticated mechanics of the Variable Reluctance motor with the power of the permanent magnet on the shaft, giving a greater torque with greater precision in the steps, which can vary between 3.6° and 0.9°, against 7.5° to 15° for permanent magnet as shown in Figure 10.13.

10.5 STEPPER MOTOR CONTROL

As previously discussed, stepper motors are driven by electrical pulses at their terminals. Such pulses must follow a suitable sequence so that the device is switched correctly and can perform the desired movement. Pay attention to the voltage and current values present in the power circuit and take care that the currents drained from the controller circuit are not high and cause the device to burn out.

There are several ways to drive a stepper motor. Some coil switching sequences will be described here.

Full-step:

B1	B2	B3	B4
1	0	0	0
0	1	0	0
0	0	1	0
0	0	0	1

Full step (highest torque):

B1	B2	B3	B4
1	1	0	0
0	1	1	0
0	0	1	1
1	0	0	1

Half step:

B1	B2	B3	B4
1	0	0	0
1	1	0	0
0	1	0	0
0	1	1	0
0	0	1	0
0	0	1	1
0	0	0	1
1	0	0	1

10.6 STEPPER MOTOR DRIVER

To drive the stepper motor, and subsequently apply the micro-step technique, it is essential to have knowledge of its driver.

The output signals of a logic sequencer are transmitted to the inputs of a power driver, through which the on/off actions on the motor windings are directed. This power driver can be called motor driver or simply driver. A fundamental circuit for applications that use a direct current motor – in the case of our study, the stepper motor – is called a bridge circuit, or "H" bridge. With it, it is possible to control the activation of the motor coil and also the direction of circulation of the electric current, which will define the side to which the motor must move.

The direction of rotation of the direct current motor depends on the polarity in which it is energized. Considering that the coil of a DC motor has two points (a) and (b) in Figure 10.14, when switch S1 is closed, the motor will be energized with the source positive at point (a). In this way, the motor will rotate clockwise.

To make the motor rotate counterclockwise, the polarity of the motor must be reversed, that is, the positive must be applied at point (b) in Figure 10.15.

Observing Figures 10.14 and 10.15, it can be deduced that to invert the direction of rotation of the motor, the polarity of the energy source must also be inverted. When there is a need to invert the direction of rotation of a motor, a circuit is set up to invert the polarity from the drive switches. Such a circuit is called an "H" bridge, whose name is associated with its construction, as can be seen in Figure 10.16.

When activating only switches S1 and S4, the motor will be polarized with positive at point "a," causing the motor to rotate clockwise. The direction of the electric current can be seen in Figure 10.17 (a).

When activating only switches S2 and S3, the motor will be polarized with the positive at point "b," causing the motor to rotate counterclockwise. The direction of the electric current can be seen in Figure 10.17 (b).

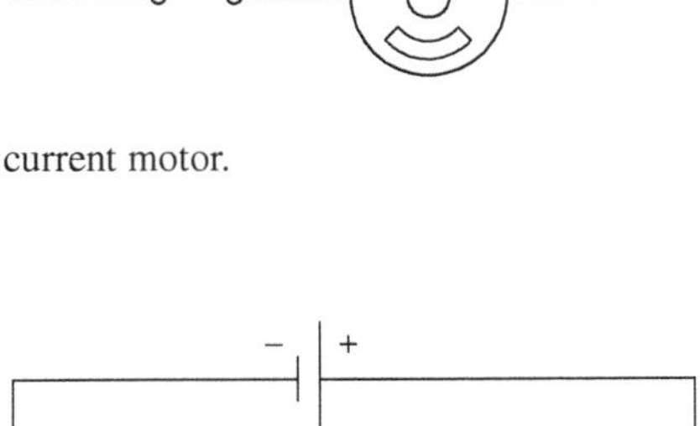

FIGURE 10.14 Direct current motor.

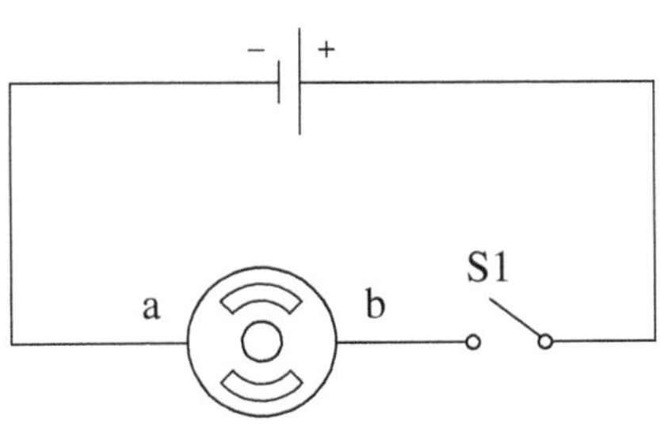

FIGURE 10.15 Direct current motor with reversed direction.

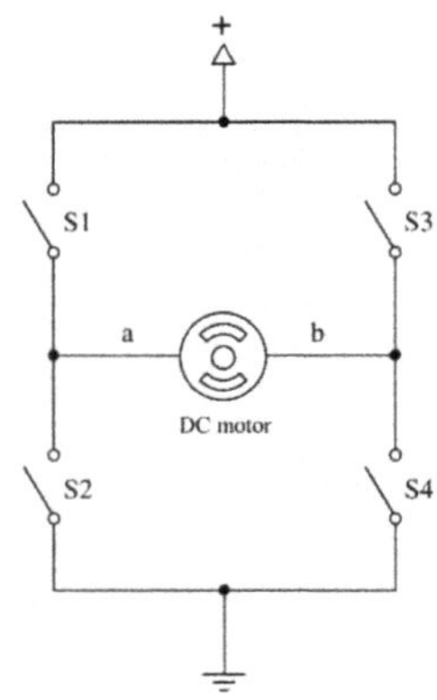

FIGURE 10.16 "H" bridge with switches.

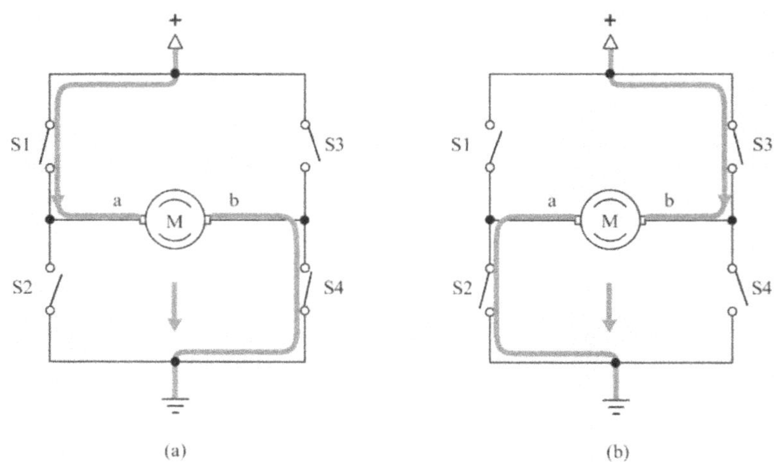

FIGURE 10.17 Motor polarization clockwise (a) and counterclockwise (b).

Replacing the switches used in the examples above by transistors, the "H" bridge circuit can be controlled by discrete signals, or even by microcontrollers.

It is important to place diodes in parallel with the switching components, in order to block reverse voltage peaks, caused when the transistors are switched off. As we can see in Figure 10.18, if transistor Q1A is conducting, the path of electric current in the circuit will be indicated by the gray trace curve that passes through the coil and conducts through transistor Q2B to ground. After cutting the transistors, the coil generates a reverse voltage that is discharged through the diodes. The path of the electric current can also be seen in Figure 10.18.

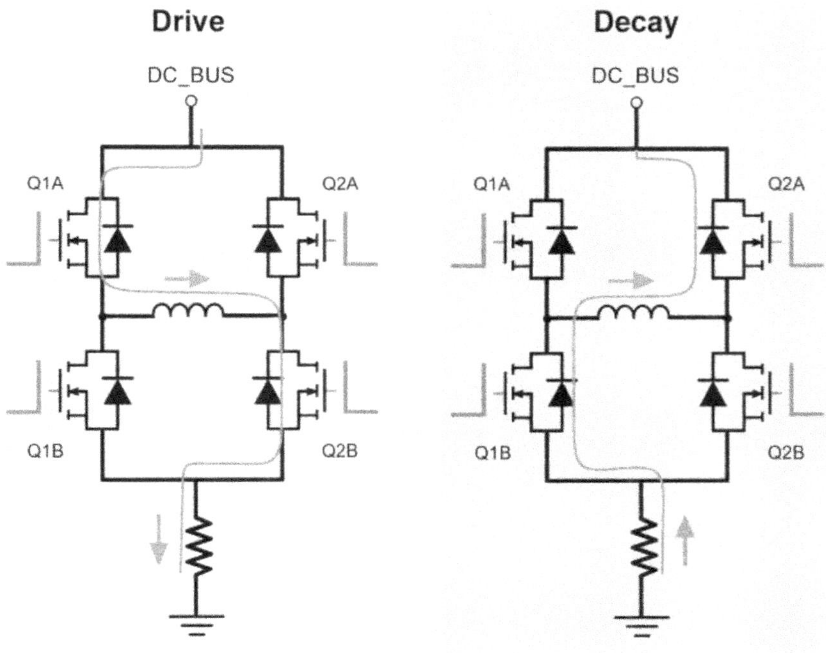

FIGURE 10.18 H bridge with transistors.

The H bridge with transistors, as illustrated in Figure 10.18, represents the standard implementation found in modern stepper motor driver integrated circuits. The Drive state allows current to flow through the motor coil in the desired direction, while the Decay state provides a controlled path for the coil's stored energy to dissipate, reducing current ripple and improving motion smoothness.

In practical applications, the switching sequence of the transistors is governed by the logic sequencer discussed earlier in this chapter. By precisely controlling the timing and polarity of each coil activation, it becomes possible to implement full-step, half-step, and micro-step driving modes. Micro-stepping, in particular, benefits greatly from the H bridge topology, as it allows intermediate current levels to be set in each coil, producing finer angular resolution and quieter operation.

Modern driver ICs integrate all of these elements — the H bridge transistors, protective diodes, and decay logic — into a single package, simplifying the design of stepper motor control systems considerably. Understanding the underlying principles of the H bridge remains essential, however, as it enables the engineer to select the appropriate driver, configure its operating parameters correctly, and diagnose faults when they arise in the field.

Suggested Reading

1. TRANSFORMERS

P. S. Bimbhra, *Electrical Machinery*, 2011.
Stephen J. Chapman, *Electric Machinery Fundamentals*, 2012.
J. B. Gupta, *Theory and Performance of Electrical Machines*, 2015.
Irving L. Kosow, *Electrical Machines and Transformers*, 1983.
V. K. Mehta & Rohit Mehta, *Principles of Electrical Machines*, 2008.
George Shultz, *Transformers and Motors*, 1980.

2. DC MACHINES

P. S. Bimbhra, *Electrical Machinery*, 2011.
Stephen J. Chapman, *Electric Machinery Fundamentals*, 2012.
A. E. Fitzgerald, C. Kingsley Jr. & S. D. Umans, *Electric Machinery*, 2002.
J. B. Gupta, *Theory and Performance of Electrical Machines*, 2015.
V. K. Mehta & Rohit Mehta, *Principles of Electrical Machines*, 2008.
Theodore Wildi, *Electrical Machines, Drives and Power Systems*, 2010.

3. INDUCTION MOTORS

D. Aliprantis & N. Bianchi, *Electrical Machines: Design and Analysis of Induction and PM Motors*, 2016.
P. S. Bimbhra, *Electrical Machinery*, 2011.
Stephen J. Chapman, *Electric Machinery Fundamentals*, 2012.
A. E. Fitzgerald et al., *Electric Machinery*, 2002.
M. G. Say, *Performance and Design of Alternating Current Machines*, 2005.
Theodore Wildi, *Electrical Machines, Drives and Power Systems*, 2010.

4. SYNCHRONOUS MACHINES

Stephen J. Chapman, *Electric Machinery Fundamentals*, 2012.
A. E. Fitzgerald et al., *Electric Machinery*, 2002.
J. B. Gupta, *Theory and Performance of Electrical Machines*, 2015.
M. G. Say, *Performance and Design of AC Machines*, 2005.
Theodore Wildi, *Electrical Machines, Drives and Power Systems*, 2010.

5. GENERAL ELECTRICAL MACHINES

P. S. Bimbhra, *Electrical Machinery*, 2011.
Stephen J. Chapman, *Electric Machinery Fundamentals*, 2012.
A. E. Fitzgerald et al., *Electric Machinery*, 2002.
J. B. Gupta, *Theory and Performance of Electrical Machines*, 2015.
V. K. Mehta & Rohit Mehta, *Principles of Electrical Machines*, 2008.
M. G. Say, *Performance and Design of AC Machines*, 2005.
Theodore Wildi, *Electrical Machines, Drives and Power Systems*, 2010.

6. ELECTRIC DRIVES & MACHINE CONTROL

B. K. Bose, *Modern Power Electronics and AC Drives*, 2002.
Austin Hughes & Bill Drury, *Electric Motors and Drives*, 2013.
Paul C. Krause et al., *Analysis of Electric Machinery and Drive Systems*, 2002.
Robert Krishnan, *PM Synchronous and Brushless DC Motor Drives*, 2010.
Werner Leonhard, *Control of Electrical Drives*, 2001.
D. W. Novotny & T. A. Lipo, *Vector Control and Dynamics of AC Drives*, 1996.

7. POWER SYSTEMS

Arthur Fitzgerald, *Electric Machinery*, 2002.
J. J. Grainger & W. D. Stevenson, *Power System Analysis*, 1994.
N. Mohan, *Electric Machines and Drives*, 2011.
Hadi Saadat, *Power System Analysis*, 2010.

8. SPECIAL MACHINES & ADVANCED ANALYSIS

B. K. Bose, *Power Electronics and AC Drives*, 2002.
R. Krishnan, *Permanent Magnet Synchronous and Brushless DC Motor Drives*, 2010.
T. J. E. Miller, *Brushless Permanent-Magnet and Reluctance Motor Drives*, 1989.
J. Pyrhönen et al., *Design of Rotating Electrical Machines*, 2013.

Index

For Product Safety Concerns and Information please contact our EU
representative GPSR@taylorandfrancis.com
Taylor & Francis Verlag GmbH, Kaufingerstraße 24, 80331 München, Germany

www.ingramcontent.com/pod-product-compliance
Ingram Content Group UK Ltd.
Pitfield, Milton Keynes, MK11 3LW, UK
UKHW021506240926
475702UK00035B/506
9781032853185